44
15123
D1880471

PLANT ANATOMY AND PHYSIOLOGY

PLANT ANATOMY AND PHYSIOLOGY

Editor
ASLAM KHAN

KALPAZ PUBLICATIONS
DELHI-110052

PLANT ANATOMY AND PHYSIOLOGY

(Botany)

© Aslam Khan

ISBN: 81-7835-049-1

All rights are reserved. No part of this book may be reproduced in any manner without written permission.

Published in 2001 in India by
Kalpaz Publications
C-30, Satyawati Nagar
Ashok Vihar Phase-III
Delhi-110052
E-Mail: kalpaz@hotmail.com

Laser Typesetting at : A.S. Printers
Printed at : Mehra Offset Printers, New Delhi

Preface

Plant Anatomy and Physiology provides a comprehensive survey of major issues at the forefront of botany. It contains a detailed study of fundamentals of plant anatomy and physiology. These information should be valuable to researchers in the field of plant development, as well as to students and professionals who want an introduction to current topics in this subject. The first part of the book deals with plant anatomy and the second part deals with plant physiology.

The first section of plant anatomy includes detailed study of the elements and tissues of which the plant is constructed and enables a better understanding of adaptation to special functions as well as of the adaptation of entire plants to different environmental conditions. Without a thorough knowledge of the anatomical and histological structure of plants the results of physiological and ecological experiments, for instance, may be incorrectly interpreted. Anatomy draws the attention of the student to the form variability and structure of the tissues comprising the plant body to develop an aesthetic sense.

The second section provides a comprehensive survey of major issues at the forefront of plant physiology. The physiological functions are discussed along with the structure in order to emphasise the nature of plants.

All the chapters in this book have been arranged in a conventional way, as it may provide good for students to pass from one to another with good continuity. Our purpose in this book is to meet the needs of teachers who present this subject to students who do not have the time or perhaps the inclination to pursue it in depth, but who may also require explicit coverage of basic topics. Teachers and scholars in botanical sciences will find this book highly valuable.

Editor

Contents

Chapter 1

General Structure of Plants

The general structures of a flower-bearing seed plant, starting with the seed, are outlined in this chapter. The seed contains an *embryo plant*, enveloped and protected by a *seed coat*, and is supplied with a source of stored food. The plant embryo contains a minute axis with two poles—the root growing point and the shoot growing point. On the minute axis occur laterally the *cotyledons* or seed leaves. The food required for the germinating plantlet may be stored in the cotyledons or in a special tissue, the *endosperm*.

Under suitable growing conditions the seed germinates and a young plant or *seedling* emerges. The seedling grows, extends its roots into the soil and its shoot (stems and leaves) into the atmosphere. The growth of the shoots and roots is due to the formation of new cells by meristematic (embryonic) tissues of the growing points, followed by growth and differentiation of these cells.

When the plant attains adult size, flowers are formed. After pollination (transfer of pollen grains from the stamens to the style) and *fertilization, fruit,* containing seeds, develop, thus completing the life cycle. Some plants die shortly after seed set (annual plants), others (perennials) continue to grow for many years and become shrubs or trees.

The plant organs, as the organs of animals, are composed of tissues (groups of cells which carry on specific activities). The cells of plant tissues are small compartments possessing living material, the protoplasm, enclosed by a cell wall. All metabolic processes take place in the cells. The roots anchor the plant in the soil, take up water and mineral salts from it, and in many cases store food. The shoot consists of stem and leaves. The leaves produce food by photosynthesis and give off water vapour by transpiration. The stem supports the leaves and has the role of conducting water and mineral salts from the roots to the leaves and synthesized organic substances from the leaves to regions of growth or storage. At the tips of the shoots and roots the apical meristems are situated. The cells that make tip these meristems divide, grow, differentiate and thus cause the extension growth of the plant. The shoot apices, together with the young developing leaves, form the buds.

The root

The main roots give off side roots which may branch again. Such a root system is termed a fibrous root system. Carrot, radish, and most other dicotyledons possess a main root which grows downwards, perpendicular to the soil surface. From this root branches arise. A root system of this type is called a *tap-root* system. Closer observations of young roots or young portions of roots reveal a region of hairs, root *hairs,* close to the root tip. These can best be seen in seedlings that have been grown on filter paper or in sand in a most atmosphere. The root hairs are cell elongations of the root epidermis and are present in great numbers. They are relatively short-lived, and new ones are constantly formed close to the root tip as the latter continues to grow and push its way through the soil. The root hairs greatly increase the surface area of the root portions which are mainly involved in water and mineral salt uptake.

Root: its internal structure

E*pidermis* is a protective tissue and consists of a single layer of densely packed cells. Below the epidermis occurs a relatively thick region, the cortex. The cortex is composed chiefly of structurally unspecialized cells, *parenchyma* cells, with large intercellular spaces. The innermost layer of the cortex is a single row of cells, the *endodermis*.

In the primary state the walls of all endodermal cells are thin except for a band-like thickening on the radial and transverse sides of the cell. These thick layer of unspecialized parenchyma cells, the *pericycle*, which originate from the same group of meristematic cells as the xylem and phloem. The pericycle which retains meristematic properties gives rise to lateral roots.

The xylem consists of conducting cells—*tracheids* and *vessel members*-as well as of *fibres* and *parenchyma*. A mature tracheid is a single elongated cell. In the thickened walls there are thin places *(pits)* through which water can easily pass. Vessel members are also single cells with walls similar to those of the tracheids but completely perforated at their ends. They are usually shorter than tracheids and are arranged in vertical rows. A row of vessel members is termed a *vessel* or *trachea* The fibres are elongated, tapering thick-walled cells functioning mainly in strengthening of tissues. The parenchyma represents a kind of filling tissue and functions in the storage of food.

The phloem consists of *sieve-tube members, companion cells, fibres,* and *parenchyma.* The sievetube members are living cells, arranged in vertical rows, which are known as *sieve tubes,* and function in the translocation primarily of organic substances. The companion cells are sister cells of the *sieve-tube* members and remain in close contact with them. In gymnosperms and dicotyledons a meristematic tissue termed the *vascular cambium* surrounds the xylem. While the apical meristem is responsible for the extension growth of the plant organs, the vascular cambium is responsible for their radial growth which brings about the thickening of the organs. The cells of the vascular cambium divide and produce xylem elements centripetally and phloem elements centrifually. The vascular tissues originating from the cambium are regarded as secondary. They are termed the *secondary xylem* and the *secondary,* phloem.

The stem

According to their external and internal characters, many types of stems can be distinguished, e.g. stems of grasses (wheat, corn), herbaceous dicotyledons (peas, sunflower, tomato), and woody stems of shrubs and trees. There are also unusual types of stem which underwent metamorphosis to special forms correlated to environment and function. To such types belong bulbs (onions) and *tubers* (potatoes) that store *food, succulent* stems of plants like cacti which store water and carry on photosynthesis, a process occurring in other plants almost exclusively in the leaves. Steins may also become modified to tendrils and thorns.

When studying the external structure of a shoot or twig it can be seen that the leaves are distributed on the stem in a specific pattern. That part of the stem to which the leaf is attached is termed the *node* and that part of the stem between two nodes, the *internode.* Buds, which are actually minute shoots, also occur on the stem. The bud terminating the shoot is called the *terminal bud.* In the axils of the leaves occur *lateral buds* or *axillary in buds* which are usually smaller than the terminal one. The buds may remain dormant for determined periods (e.g. during the winter in deciduous trees) or for many years (most of the axillary buds).

As long as the cells of the apical meristem of the terminal bud retain the ability to divide, most of the axillary buds remain dormant and the shoots which develop from axillary buds are limited in growth. The extent of growth of the side branches and the dormancy of the axillary buds is controlled by the terminal bud. In plants in which the terminal bud of the main axis remains active throughout the life of the plant, the branching of the stem is *monopodial.*

In many plants the apical meristem of the terminal bud, after continuing for some time to contribute to the elongation of the stem and to producing leaves, may abort or become reproductive and produce a flower. In both these cases further growth is carried out by lateral buds, and the side branches take over the function of the main primary axis. Such branching is termed *sympodial.*

Stem: its internal structure

When examining young stem portions with the aid of a microscope the following arrangement of tissues from the outside towards the centre of the stem can be distinguished. The *epidermis* is coated by a very resistant substance of lipold nature called *cutin*. This superficial layer of cutin is coined cuticle. It is, especially when thick impermeable to water and gases. Epidermal appendages such as hairs are often present.

The zone between the epidermis and the vascular tissue is termed the cortex. In the cortex three types of tissues may occur: *parenchyma, collenchyma,* and *sclerenchyma.* The *parenchyma* may store food and in green stems also produce food by photosynthesis. The *collenchyma* and sclerenchyma serve its strengthening tissues. The collenchyma cells have unlignified walls which are thickened usually in the corners of the cell. The sclerenchyma cells have very thick, mostly lignified walls. Two types of sclerenchyma cells can be distinguished: fibres, which are long cells, tapering at their ends, and *sclereids*, which are more or less isodiametric or branched. Inside the cortex is the vascular cylinder, which contains the vascular tissues. The vascular tissues are generally organized—in the young stem portions—in strands called *vascular bundles.* Each bundle contains xylem on the inside facing the centre of the stem and phloem on the outside, In some cases, e.g. in pumpkin, phloem appears on both sides of the xylem.

In the gymnosperms and dicotyledons the vascular bundles are arranged in a circle around a pith, which is usually parenchymatous, whereas in monocotyledons they are generally scattered throughout the stein cross-section. Another difference between the monocotyledons and the dicotyledons and gymnosperms is the appearance of a vascular cambium between the xylem and phloem only in the vascular bundles of the latter. The vascular cambium, which appears distant from the apices, produces additional xylem centripetally in continuation to the xylem of the vascular bundles and additional phloem centrifugally, thus pushing to the outside the phloem of the bundles. The tissues of the vascular bundles are produced directly by the apical meristems and are called *primary xylem* and *primary phloem.* Those produced by the vascular cambium are termed *secondary xylem* and *secondary phloem.*

Cambium also develops between the primary vascular bundles, thus forming a whole ring of cambium, as seen in cross-section. From this stage the cambium starts to produce a solid mass of secondary xylem or *wood* which in some trees may reach several metres in diameter. The secondary phloem is usually produced in smaller quantities and, together with remnants of the epidermis, cortex and additional tissues, which will be discussed later, it forms the *bark.*

The secondary xylem or wood consists basically of the same elements as the primary xylem but is usually richer in fibres and is penetrated radially by parenchymatous rays called vascular rays. In trees with seasonal growth cycles the wood exhibits rings when cross-sectioned. This is due to the denser structure of the xylem produced towards the end of the growth season.

While the cambium causes increase in width of the stem and root by adding more and more wood increments, extension growth of the young axes continues by the activity of the apical meristems. With the increase in thickness of the wood, pressure is imposed on the epidermis and cortex which eventually rupture.

A secondary protective tissue—the *cork layer* or *phellem*—then takes over the primary function of the epidermis. The cork is composed of dead flattened cells with no intercellular spaces. Its cells are lined by a lamella of a fatty substance, called *suberin.* This tissue is produced by the *cork cambium* or *phellogen* which, in the stems of most plants, arises in the outer cortical cells, and in the roots generally in the pericycle. In most long living plants, with the continuous increase in diameter of the axis and bursting of the first formed cork, successive cork cambia develop progressively further inwards, i.e. in the phloem.

Since cork is an impermeable barrier, when formed in the secondary phloem it causes the death of that part of the phloem which is situated on its outside. All the tissues, from the vascular cambium outwards, form the bark of the axis. The cork divides the bark into the living *inner hark* and the dead *outer bark*. Oxygen supply to the living tissues which are covered by cork is effected by groups of cells with intercellular spaces called *lenticels.*

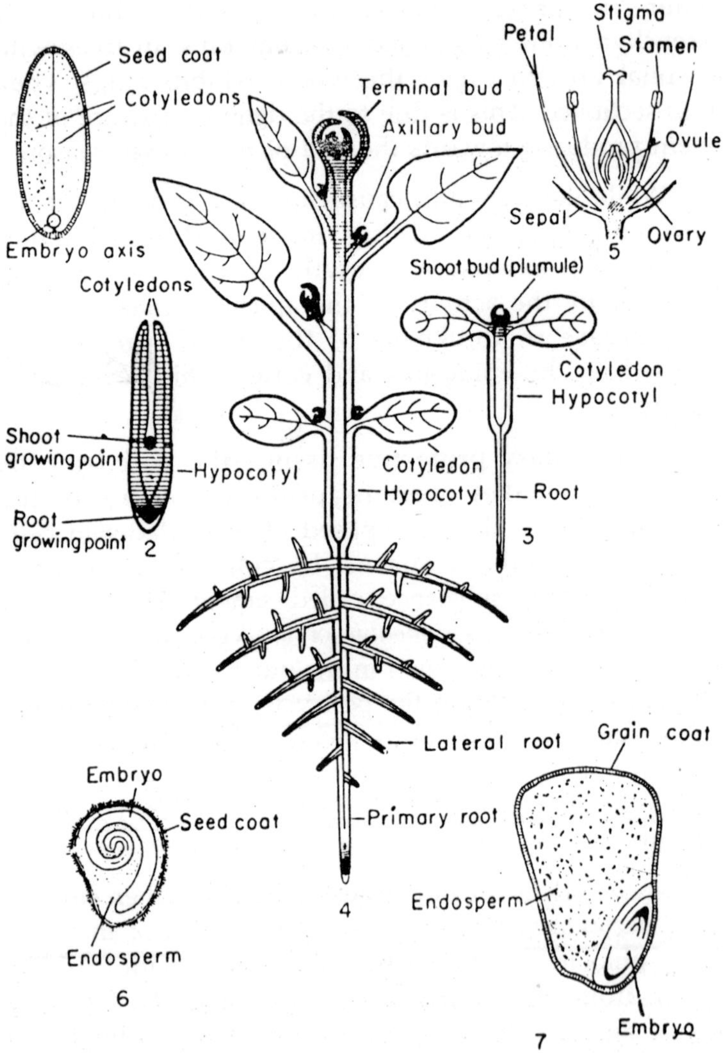

Fig. 1. Schematic drawing of longitudinal sections of various plant organs

The leaf

Leaves synthesis organic compounds using light as a source of the necessary energy, a process known as *photosynthesis.* This energy-converting process takes place within special cell organelles called *chloro*plasts, in which the pigment *chlorophyll is* present. The external and internal structure of the leaf is correlated to its role in photosynthesis and transpiration.

The leaf is flat and thin, thus enabling the solar rays to penetrate into all its cells. The high surface to volume ratio also enables successful gas exchange. The veins seen in the *leaf blade (the* expanded part of the leaf) contain vascular tissues. In the parenchymatous tissue (the *mesophyll)* occurring between the upper and lower epidermis of the leaf two zones can be distinguished: the *upper—palisade* parenchyma—consisting of elongated cells, and the lower-spongy parenchyma—consisting of irregularly shaped cells with large intercellular spaces.

The palisade parenchyma is more densely packed with chloroplasts. In the epidermis occur stomata, which serve for gas exchange between the leaf tissues and the atmosphere. Each stoma consists of two guard cells surrounding an aperture. The stomata may open or close the aperture, and thus regulate the entrance and exit of gases to and from the leaf.

The flower

Flowers are the reproductive organs of palnts. The flower is formed by a specialized apical meristem which has developed from a vegetative shoot apex after being induced to do so by internal or external factors. The flower consists of a group of specialized leaves: sepals, petals, stamens, and carpels. All the sepals, which are usually green, constitute the calyx; all the petals, which are usually coloured and attractive, together constitute the corolla. Each stamen consists of a stalk—the filament-bearing an *anther in* which pollen grains develop. The pollen grains contain male *gametes or* sperm cells. The carpels occur singly or in groups, and they comprise the pistil in which three parts can be distinguished: the basal part—the ovary; the slender middle part—the style; and the top part—the stigma.

The ovary bears in its locule or locules the ovules in which the female gametes—the egg cells—develop. The pollen grains are transferred by wind or insects from the ripe anther to the stigma of the pistil. This process is called *pollination.* The pollen grain germinates on the stigma to form a *pollen tube,* which contains two sperm cells. When the pollen tube penetrates the ovule, one of the sperm *cells fertilizes the* egg cell and a zygote is formed.

Development of meristems

During early stages of development of the embryo all the cells undergo division, but with further growth and development cell division and multiplication become restricted to special parts of the plant which exhibit very little differentiation and in which the tissues remain embryonic in character and the cells retain the ability to divide. These embryonic tissues in the mature plant body are called *meristems.*

Cell division can also occur in tissues other than meristems, for instance in the cortex of the stem and in young, developing vascular tissues. However, in these tissues the number of divisions is limited. On the other hand, the cells of the meristems continue to divide indefinitely and as a result new cells are continually added to the plant body. Meristems may also be found in a temporary resting phase, for instance in perennial plants that are dormant in certain seasons and in axillary buds that may be dormant even during the active phase of the plant.

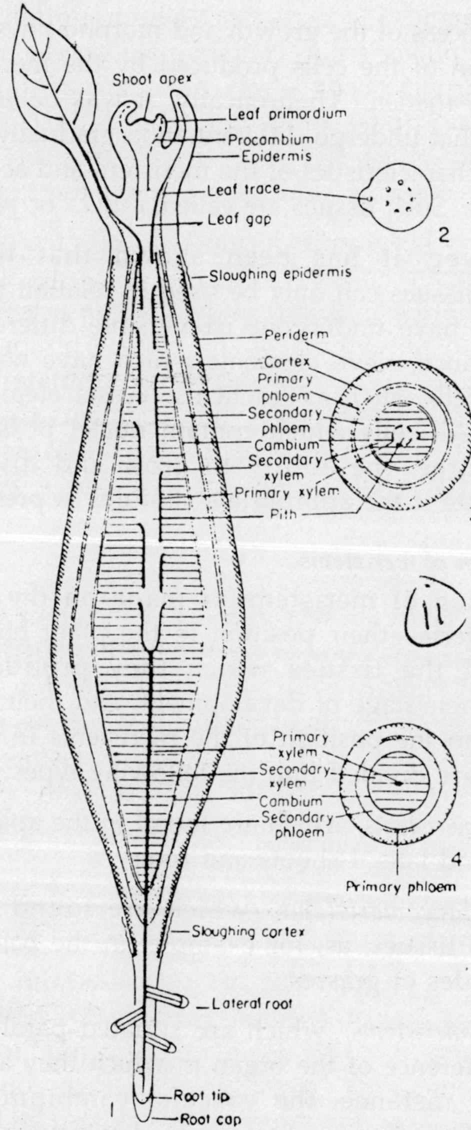

Fig. 2. Schematic drawing of dicotyledonous plant showing arrangement of principal tissues 1. Longitudinal section, 2, 3, and 4 Cross sections at various levels

The process of the growth and morpho-physiological specialization of the cells produced by the meristems is called *differentiation.* Theoretically, it was believed that the tissues that undergo differentiation gradually lose the embryonic characteristics of the meristem and acquire the mature state. Such tissues are called *mature* or *permanent.*

However, it has been shown that the term permanent tissues can only be used in relation to certain cells which have undergone irreversible differentiation, as, for instance, sieve elements which have no nucleus and dead cells, such as tracheids, vessel elements and cork cells. All cells which contain nuclei possess, to a certain degree, the ability to grow and divide and redifferentiate if the appropriate stimulus is present.

Classification of meristems

Classification of meristems is made on the basis of various criteria—their position in the plant body, their origin and the tissues which they produce, their structure, their stage of development, and their function. According to the position of the meristems in the plant body they are divided into the following types:

i) *Apical meristems,* which are found in the apices of the main and lateral shoots and roots,

ii) *Intercalary meristems,* which are found between mature tissues, as, for example, in the bases of the internodes of grasses;

iii) *Lateral meristems,* which are situated parallel to the circumference of the organ in which they are found, as, for instance, the vascular cambium and the phellogen. It is customary to distinguish between *primary* and *secondary meristems*—a classification based on the origin of the meristems.

Primary meristems are those whose cells develop directly from the embryonic cells and so constitute a direct continuation of the embryo, while secondary meristems are those that develop from mature tissues which have already undergone differentiation. The above definitions of primary and secondary meristems, however, are not always accurate.

For example, the apical meristems of truly adventitious organs develop secondarily within relatively mature tissues as well as within secondary meristematic tissues, although according to their structure and function they are primary meristems. On the other hand, a large part, or sometimes even the whole, of the vascular cambium, which is generally accepted to be a secondary meristem, develops, at a late stage, from the apical meristem, i.e. from a part of the procambium.

Examples of secondary meristems, which can be determined as such without doubt, according to origin, are the phellogen which develops from parenchyma or collenchyma cells which have already undergone differentiation and callous tissue which develops in tissue cultures made from mature tissues.

It is more correct to use the terms primary and secondary meristems to refer to the stage of development at which the meristems appear, and to the types of tissue that develop from them, and not to their origin. From the primary meristems the fundamental parts of the plant, such as the epidermis, the cortical tissues of the stem and root, the mesophyll of the leaf, and the primary vascular tissues, develop, and from the secondary meristems the secondary vascular and protective tissues.

In certain monocotyledons, such as some palms, banana, *Veratrum,* and others, the thickening of the stem takes place near the apices and therefore is regarded as being of primary nature. The meristem responsible for this type of increase in thickness is termed *primary thickening* meristem.

Development of primary meristems

In the primary meristems, however, different regions in various stages of differentiation can be distinguished. In primary apical meristems we distinguish a *promeristem* and a meristematic zone below it in which groups of cells have undergone a certain degree of differentiation. The promeristem consists of the *apical initials* together with the cells derived from them and which are still close to the initials.

The partly differentiated meristematic zone consists of the following three meristems: the *protoderm* from which the epidermal system of the plant develops, the *procambium* from which the primary vascular tissues develop, and the *ground meristem* from which the ground tissues of the plant, as, for instance, the parenchyma and sclerenchyma of the cortex and pith and the collenchyma of the cortex, develop. The term initials in meristems refers to cells which always remain within the meristem. When an initial divides one of the daughter cells continues to fulfil the original function of an initial, whereas the other daughter cell, after several divisions, undergoes differentiation and maturation. Newman refers to an initial in the meristem as a *continuing meristematic residue.*

Meristematic cells are usually thin-walled, more isodiametric in shape than the cells of mature tissues, and relatively richer in protoplasm. However, it is not possible to find a general morphological criterion by which meristematic cells can be distinguished from unspecialized mature cells. Usually the protoplasts of meristematic cells are devoid of reserve materials and crystals, and the plastids are in the proplastid stage.

However, the protoplasts of the phellogen, a secondary meristem, may contain these bodies.

In most cells of apical meristems of a large number of plants, and especially among the angiosperms, the vacuoles are very small, not obvious, and are scattered throughout the protoplast. However, in pteridophytes and many spermatophytes at least some of the cells of the apical meristem contain conspicuous vacuoles. Also the cells of the vascular cambium are highly vacuolated.

In general, it is possible to state that the larger the meristematic cell, even if it is an initial cell or one close to the initials, the greater is the degree of vacuolization. The size of the meristematic cells varies. Also the ratio between the size of the cell and that of the nucleus varies very greatly in different meristematic cells. The wall of meristematic cells is usually thin, but certain cells in the apical meristems have thick walls, and cells of the vascular cambium have very thick radial walls at certain periods. From the above it can be seen that morphological analysis alone is not sufficient to determine the meristematic nature of cells and the use of experimental methods is often necessary.

Apical meristems

The apical meristem will be divided, as already mentioned, into two main regions—the promeristem, which comprises the apical initials and neighbouring cells, and the meristematic zone behind it in which the three basic meristems of the tissue systems can be distinguished. The following discussion deals mainly with the arrangement and function of the cells in the promeristem

Shoot apex

The term *shoot apex* is generally used as it is the region of initiation of the primary organization of the shoot in which the processes of growth take place and which cannot be limited to a point. The shoot apex proper is considered as that terminal part of the shoot immediately above the uppermost leaf primordium. There are great differences in the shape and size of the shoot apices among the spermatophytes. In a median longitudinal section the apex generally appears more or less convex.

In *Anacharis* and *Myriophyllum* and some grasses the shape of the apex is a narrow cone with a rounded tip, while in a few plants, e.g. *Drimys* and *Hibiscus syriacus,* it is slightly concave. Before the initiation of each leaf the apical meristem widens considerably and after the appearance of the leaf primordium it again becomes narrow. This phenomenon is rhythmic, i.e. it recurs with the initiation of each leaf or pair of leaves. Schmidt introduced the terms minimal and maximal areas of the apex. For the period between the successive initiations of two leaves or two pairs of leaves he suggested the use of the term *plastochron* which had been used previously but with a much wider meaning.

The shoot apices of dicotyledons with opposite leaves (such as *Lonicera, Coleus, Vinca, Ligustrum, Syringa,* and others) are particularly suitable for the study of plastochronic changes. Successive plastochrons, especially of genetically uniform plants grown in a controlled environment, may, at least during part of their vegetative growth, be of equal duration.

In the Angiospermae the shoot apices are usually small. The measurement of the diameter is taken as the width of the apex immediately above the youngest primordium. The diameter usually varies between 90 μ (in certain grasses) and 130-200 μ (in many dicotyledons). In the banana plant, however, the width of the apex reaches 280 μ, in certain Palmae and in *Nymphaea,* 500 μ, in *Trichocereus* it is between 700-800 μ and in *Xanthorrhoea media* the maximum diameter reported was 1283 μ. The differences in the diameter of the apices of gymnosperms are much greater. The apices of the conifers are cone-shaped and fairly narrow and the dimensions of their diameter are similar to those typical for angiosperms.

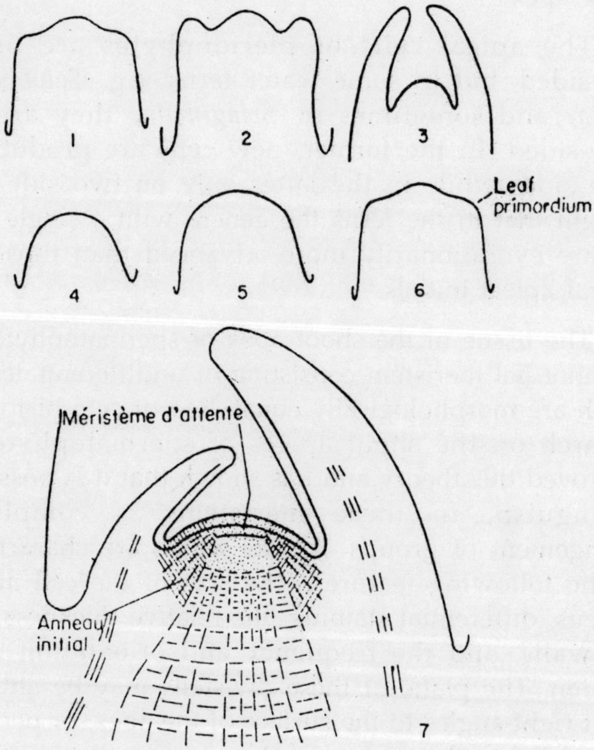

Fig. 3. Diagrams of longitudinal sections of a shoot apex

In Pteridophyta there are one or more initials which can usually be easily distinguished from the neighbouring cells. These initials give rise to all the cells of the apex. If only one initial is present it is termed the *apical cell* and if more than one cell are present they are termed *apical initials.* The single apical cell usually is tetrahedral in shape and its base is directed towards the surface of the apex. A single apical cell is found in the Psilotales, in *Equisetum* and in some ferns. The single apical cell divides in such a manner that the new cells are formed on all its sides with the exception of that on the surface of the apex.

The apical cells of pteridophytes are usually four-sided, but in some water ferns, e.g. *Salvinia* and *Azolla,* and sometimes in *Selaginella,* they are only three-sided. In the former, new cells are produced on three sides while in the latter only on two sides. It is thought that in the ferns the genera with a single apical cell are evolutionarily more advanced than those with several apical initials.

The tissue of the shoot apex of spermatophytes was a primordial meristem consisting of undifferentiated cells which are morphologically equal. Recent cyto-histological research on the shoot apices of spermatophytes has disproved this theory and has shown that it is possible to distinguish, in these meristems, a complicated arrangement of groups of cells which are characterized by the following features—the size of the cell and the nucleus, differential staining, the relative thickness of the cell walls, and the frequency and orientation of cell division. The plane of these divisions may be anticlinal, i.e. at right-angles to the surface of the apex, or periclinal, i.e. parallel to the surface of the apex, or diagonal.

It is characteristic of all the gymnosperms that the direction of the cell divisions in the surface of the apex is both anticlinal and periclinal, and so the top layer represents the initiation zone of the entire apex and has been termed the *surface meristem.* In the cells of the summit of this meristem the periclinal divisions are more frequent. These cells are the *apical initials.* The striking feature in the structure of most gymnosperm apices is the occurrence of a distinct zone of *central mother* cells, which occurs in a median position below the surface layer. The central mother cells are relatively large, polyhedral, irregularly arranged, and their cell walls are thick, particularly in the angles of the cells.

The apical initials and the central mother cells contain numerous vacuoles and large light-staining nuclei. Cecich who has studied the shoot apical meristem of *Pinus banksiana* with the aid of the electron microscope has observed that the apical initials and the central mother cells each contain numerous lipid bodies and their nuclei contain very little, if any, heterochromatin. The vacuoles seen in the apical initials and central mother cells in light microscopical sections, prepared with standard histological techniques, represent sites of dissoluted lipid bodies. Thus the lack of heterochromatin in the nuclei and the dissolution of lipids in the cytoplasm appear to contribute most to the appearance of the cytohistological zonation of the shoot apex.

Along the sides and the base of the central mother cell zone the other apical regions develop as a result of the diagonal and horizontal divisions of the central mother cells. In this way the *peripheral* or *flank meristem* is developed laterally and the *rib meristem* zone from the base. The peripheral meristem forms a kind of a short cylinder surrounding the rib meristem. The term *rib meristem* was introduced by Schuepp to describe that type of meristematic tissue that consists of vertical series of transversely dividing cells. The cytoplasm of the peripheral zone is dense. According to Cecich it contains numerous ribosomes. The cytoplasm of the rib meristem is less dense and contains numerous small vacuoles. According to Popham, three principal types of gymnosperms can be distinguished on the basis of the structure of the shoot apex.

The Cycas type of root apex lacks the zone of central mother cells. Here three meristematic zones can be distinguished.

i) The *surface meristem* in which the cells divide anticlinally, periclinally, and diagonally. The cells of this zone are not uniform in appearance, and apical initials have been distinguished in the centre of this zone in the seedlings of *Cycas revoluta,* but not in mature plants. The cells of this zone give rise to the epidermis and the other apical meristematic zones.

ii) The *rib meristem* which is situated in the central region of the apex below the surface layer. In the upper region of this zone vertical rows of cells are obvious. The cells at the base of these rows divide periclinally, anticlinally, and diagonally, and they are usually large and contain large vacuoles.

iii) The *peripheral meristem* which enlarges by cell division within the zone itself and by the addition of cells from the surface layer and from the periphery of the rib meristem. The cells of this zone are smaller than those of the rib meristem and they are generally elongated.

In the *Ginkgo* type five meristematic zones can be distinguished in the apex.

i) The *surface meristem;*

ii) The zone of *central mother cells;*

iii) The *rib meristem* from which the pith of the stem develops;

iv) The *peripheral meristem,* and in addition, there is a cup shaped zone termed *cambium-like transitional zone.* This zone forms a transitional zone between the central mother cells and the rib and flank meristems, and is characterized by frequent cell divisions.

Most of the divisions are periclinal in relation to the central mother cells and so cells are added to the zones below, i.e. to the flank and rib meristems. The number of cells in the flank meristem increases by division in the meristem itself as well as by the addition of cells from the surface meristem and the cambium-like transitional zone. The cortex, leaf primordia, procambium and, in certain plants (*Microcycas* and *Zamia*), the outer region of the pith develop from this zone. T

In the *Cryptomeria-Abies* type of shoot apec four meristematic zones can be distinguished. The cambium-like transitional zone is absent and the remaining zones are as in the *Ginkgo* type. Of the plants with this type of apical meristem the following species should be mentioned: *Pinus montana, Sequoia gigantea, Metasequoia glyptostroboides, Abies concolor, Taxus baccata, Ephedra altissima,* and *Cryptomeria japonica.*

At the beginning of the cyto-histological research on the apices of plants the *histogen theory* of Hanstein was put forward. According to Hanstein the following three zones can be distinguished in the shoot apex of angiosperms: an outermost zone, the *dermatogen;* a central zone, *plerome,* which consists of irregularly arranged cells; and a hollow cylindrical zone of several layers of cells between the dermatogen and the plerome, the *periblem.* Hanstein stated that the dermatogen, periblem and plerome develop from independent groups of initials, which act as direct histogens.

According to this theory, therefore, the meristems are destined from the beginning to produce certain tissues, i.e. the epidermis develops from the dermatogen, the cortex, and internal tissues of the leaf from the periblem, and the central cylinder from the plerome. The histogen theory of Hanstein was accepted for a long time, but later research, which brought to light the following facts, disproved it. (1) In most spermatophytes it is not possible to distinguish clearly between the periblem and plerome. (2) No predetermination of the mature tissues can be traced in the various initials.

In 1924 the theory of Schmidt, which divides the apex into two regions, the *tunica* and the *corpus,* was postulated. According to this theory no constant relationship can be traced between the particular initials of the promeristem and the inner tissues of the shoot. The two regions recognized by this theory are usually distinguished by the planes of cell divisions in them. The tunica consists of the outermost layer or layers of cells which surround the inner cell mass—the corpus, The plane of cell division in the tunica is principally anticlinal. In the corpus the planes of cell division are in all directions.

Recent electron microscope studies on shoot apices revealed that the ultrastructural differences between tunica and corpus cells are mainly quantitative. The tunica enlarges in surface area and the corpus in volume. The *tunica-corpus theory* is a very adaptable one and today it is generally accepted in literature. The number of layers is not always constant in a given genus or family, and not even in a species, and cases are known where the number of layers varies in a single plant during different stages of the development of the vegetative apex. The number of layers usually varies between one and nine. In *Xanthorrhoea media* Staff distinguished as many as 10 or sometimes even 18 layers.

Sometimes, especially among the monocotyledons, a few periclinal divisions occur in the tunica, which is a contradiction of the original definition of the tunica and therefore Popham and Chan introduced the term *mantle* for all the outer layers of apex which can be distinguished histologically from the inner cell mass, the *core,* without taking into account the planes or division in these layers. They retained the term tunica for those layers in which only anticlinal cell divisions take place. In this book the term tunica is used in the broad sense, and is equivalent to the mantle of Popham and Chan. One zone is the central apical zone consisting of one or few *initials* which are usually larger and have larger nuclei and vacuoles than the other tunica cells and which, therefore, are also more lightly staining. The second zone is that region on the sides of the apex between the initials and the leaf primordia. It consists of smaller, more darkly staining cells which divide more frequently and among which periclinal division may occur close to the primordia. The corpus is far less homogeneous than the tunica.

Among the angiosperms two main types of corpus are recognized on the basis of internal arrangement. The usual *angiosperm type* in which three main zones can be distinguished in the corpus: (a) the zone of *central mother cells,* which represent the corpus initials, located below the apical portion of the tunica, i.e. below the tunica initials; (b) the *rib meristem;* and (c) the *peripheral meristem.* The latter two zones appear as continuations of the central mother cells. The *Opuntia type* in which, in addition to the above zones, a *cambium-like transitional zone* can be distinguished. This zone, which is cup-shaped, is found between the central mother cells and the rib and peripheral meristems. The cambium-like transitional zone differs from the other zones of the apical meristem in that its height and diameter vary considerably during the plastochron, reaching a maximum development close to a developing primordium.

According to Philipson this zone is only a temporary feature in many of the plants in which it occurs, as it disappears towards the end of the plastochron. The following are some examples of plants with an *Opuntia* type apex: *Phoenix dactylifera,* the Dwarf Cavendish banana, *Chrysanthemum morifolium, Opuntia cylindrica, Bellis perennis, Xanthium pennsylvanicum. Liriodendron tulipifera* and *Bougainvillea* spectabilis. In some of these species the cambium-like zone can be seen in certain phases of the plastochron only, whereas in others it appears to occur consistently.

It seems therefore preferable not to use this zone as a criterion for classifying shoot apices of the angiosperms into two different types. The cells of the peripheral meristem are usually derived from the initials of the corpus, but in some plants they-arise also from the tunica. The rib meristem usually consists of cells arranged in rows which narrow towards the apex but they may sometimes be irregularly arranged. The majority of divisions in this zone are horizontal, but diagonal divisions also occur. Series of cells that increase in size as they become further distant from the central mother cells can be distinguished. With the increased cell size the vacuoles also enlarge. From what is known today it can be concluded that the epidermis and its derivatives originate from the outermost layer of the tunica. The peripheral meristem contributes to the development of the leaf primordia, the cortex, all or part of the procambium, and sometimes also the outer region of the pith.

From the point of view of activity two zones, which are parallel to the cyto-histological zones, are generally distinguished:

i) A central apical zone which includes the initials of the tunica, and of the corpus and in which division is considered to occur rarely, and

ii) A peripheral zone to which much mitotic activity is ascribed.

Extreme views as to the activity of these two zones are held by Buvat and other French workers who, contrary to most investigators, do not regard the central apical zone as having the role of cell-producing-cells during the vegetative development of the plant, but as being active only when the vegetative apex becomes reproductive. This view is based on Plantefol's theory on phyllotaxis which attributes the organogenic and histogenic role only to a lateral, subterminal zone, the *anneau initial* (initiating ring) at which the leaf parastichies (helices) end. The mitotically inactive central apical zone was termed *meristeme d'attente* (waiting meristem). This view was strongly supported by Loiseau and Nougarede.

However, the findings of most other investigators contradict this theory of quiescence of the central apical zone. The results of the study of natural and colchicine-induced polyploid chimeras favoured-the concept that divisions do occur in this zone. The investigations on variegated chimeras lead to a similar conclusion. From histological investigations made on apical meristems of many species, cell divisions in the zone of the central mother cells were demonstrated. With the use of microphotographs taken with a cine camera of the surfaces of culture-grown shoot apices of several plants. Ball came to the conclusion that all the cells in the surface layer, including those at the summit of the apex, divide and do so quite frequently. Also the application of radioactively labelled precursors of DNA synthesis did not reveal the existence of an inactive zone in the shoot apex. In later accounts the supporters of the *meristeme d'attente* theory agreed that a few mitoses may occur in the central apical zone but claim that their role is minimal in relation to that played by the *anneau initial.*

West and Gunckel, who made cyto-histochemical studies of the shoot tips of *Brachychiton* species, found no great difference in RNA synthesis between the central apical cells and the cells of the flanking zone. Denne, who studied the duration of the mitotic cycle in the vegetative shoot apex of *Trifolium*, reports of relatively small (± 3:4) differences between the above zones. In young embryo shoots of *Zea mays*, Clowes found recently that the rates of division of cells at the summit of the apex do not differ greatly from those in the rest of the apex. Lyndon summarizing data on relative cell division rates in the various zones of the shoot apex concluded that in all or most plants there exists a gradient in the rate of division and growth from a minimum at the summit of the apex to a maximum in the region of leaf initiation. The view expressed by some authors that the apical cell of the pteridophytes serves as a unicellular quiescent centre, without histogenic potential, has also been contradicted by Polito, who used the colchicine-induced metaphase accumulation technique to show that this cell can be active. The question, however, is a quantitative one, as divisions also occur in the quiescent centre of spermatophytes, especially in young roots.

Branching of shoot

Branching of the shoot originates at the shoot apex. In many pteridophytes and in a few angiosperms the branching is brought about by the equal division into two of the single apical cell or group of apical cells as a result of which two apices are formed. This type of branching is termed *dichotomous branching*. In other pteridophytes and in most spermatophytes, lateral axillary buds, from which branches may develop later, are initiated in the axils of the leaf primordia. Accessory buds originate from sectors of the peripheral meristem of the axillary buds. Some branches may develop extra-axillary, e.g. the tendrils of the Vitaceae.

Reproductive apex

According to the-classical concept, traced to the writings of Goethe, the flowering shoot is homologous to the vegetative shoot. Gregoire, on the basis of histogenetical studies, opposed the theory that the flower is a metamorphosized shoot. According to Gregoire the reproductive meristem possesses a structure fundamentally different from the vegetative meristem. A meristematic mantle is present in the apices of reproductive axes, but it is not formed from a transformation of the tunicacorpus layers present in the vegetative apex. The two superficial layers of the mantle give rise to the floral parts and contribute cells to a parenchymatous core. Gregoire's concept did not receive much support.

Most investigators criticized it and concluded that at the time of flowering the vegetative shoot apex, terminal or lateral, undergoes various physiological and histological changes and becomes directly transformed into a reproductive apex. This may develop either a flower or an inflorescence. There are inflorescences and even single flowers which, similarly to vegetative shoots, also exhibit plastochronic size fluctuations in the floral apex during successive stages of development of bracts, perianth, stamens, and carpels. Some reproductive apices may retain the cytohistological zonation of the vegetative apex, although their form may vary. Plantefol considered that the reproductive apex originates by progressive metamorphosis of the vegetative apex. According to Plantefol the *anneau initial* functions during vegetative growth in production of foliage leaves. It forms also the sepals and even the petals. Stamens and carpels derive from the *meristeme d'attente*.

However, most authors do not support the view of the existence of a "waiting meristem" which becomes active only when the apex becomes reproductive; they consider the apex as a whole to enter a new phase of development. Philipson stated that the basic function of the vegetative apex is to promote longitudinal growth of the axis, while that of the reproductive apex is to produce a meristematic envelope with a large surface area from which the parts of a flower or flowers develop. Inside this meristematic envelope there is a rib meristem consisting of relatively large, vacuolated cells.

Many investigators have shown that the transition from vegetative to reproductive apex is gradual. Gradually this activity spreads into the central mother cell zone where the cells then become smaller and richer in protoplasm. In this way all the cells above the rib meristem are added to the tunica, the cells of which are more or less isodiametric and are relatively small. Following these changes, mitotic activity and growth ceases, or almost so, in the cells of the rib meristem and of the pith below it. Thus in the apex a parenchymatous pith surrounded by meristematic cells develops. Depending on the species, only the flower parts or the bracts, the axillary branches of the inflorescence, and the flowers themselves develop from these meristematic cells.

Many cyto-histological studies on the transition of the vegetative apex to reproductive state have been made. During transition to flowering a general increase in concentration of cytoplasmic basic proteins (historic), RNA, and total protein was found to occur in all zones of the apex.Following is a summary of events occurring in the apical meristem during flower induction in *Sinapis alba* by transfer from short to long days, as described by Bernier et *al.*:

i) Rise in the mitotic index (percentage of nuclei observed to divide) culminating about 30 h after beginning of the long day;

ii) A stimulation of DNA synthesis reaching a maximum at the 38th hour;

iii) Increase in a nucleolar diameter rising to a maximum at about the 54th hour;

iv) Increase in cell volume culminating at the 62nd hour.

v) Second rise in the mitotic index rising to a maximum at 62 h.

Floral induction was suggested to occur during the first rise in mitotic activity. Flower buds were initiated in co-ordination with the second rise in mitotic activity. The cyto-histological changes in the apex during induction of flowering recently described by many authors are no doubt preceded and accompanied by physiological and biochemical changes. One of these changes may be demonstrated by the fact that the dominance of the main apex, which suppresses the development of the lateral buds, is lost with the production of the inflorescence. An apex in this stage of development ceases to elongate in plants with capitula or single flowers, and in other plants the rate of elongation is reduced. In certain plants, such as banana and pineapple, however, marked elongation takes place.

Root apex

The structure of the root promeristem has been intensively studied in order to discover the origin of the various tissues. In some of the Pteridophyta, as, for instance, the Polypodiaceae, Ophioglossaceae and *Equisetum*, the entire root develops from a single apical cell, while in others, e.g. the Marattiaceae, a few initials are present.

When there is only a single apical cell it is tetrahedral, and it divides in such a way as to add new cells to the body of the root from its upper three sides and to the root cap from its base. In some pteridophytes, e.g. *Azolla*, after formation of the apical cell a division parallel to the distal face produces the precursor cell of the root cap. All subsequent divisions of the apical cell occur at the three proximal faces. Active division of the apical cell is confined to the early stage of root development.

At later stages of development the apical cell becomes progressively more vacuolate and ceases to divide. According to various investigators it appears that there is a single central initial or but a few initials in the root apex of the Spermatophyta.

Other investigators, however, such as Clowes, believe that there is a larger group of initials in the median region of the root apex. Allen, working on *Pseudotsuga*, distinguished a central group of permanent initials with three groups of temporary initials (mother cells) on its periphery. He observed that the meristem of the vascular cylinder developed from the first group of temporary initials, the meristem of the cortex from the second, and the *columella* from the third. The columella is a group of cells that forms the longitudinal axis of the root cap. In it the cells are arranged in longitudinal rows. Cells are added to the root cap from the columella by periclinal cell division on its periphery. The protoderm was seen to develop from the young cortex.

According to Wilcox's work on *Abies procera* there appear to be two groups of temporary initials, one of which gives rise to the central cylinder and the other to the columella, from which the root cap and cortex develop. Although research on the development of the histogens in the shoot apex has proved that they do not exist, many authors still use the terms *dermatogen*—meristem of the epidermis, *periblem*—meristem of the cortex, and *plerome*—meristem of the central cylinder, in connection with roots. However, the terms, as used today, have a meaning somewhat different from those as used by Hanstein. The mother cells of the various tissue systems of the root are replaced, at relatively long intervals, by new cells which are derived from the common permanent initials.

In many cases more than one tissue develops from a group of temporary initials and so it is desirable to use, wherever possible, instead of histogens, the terms *protoderm, meristem of the cortex,* and *meristem of the vascular cylinder* for the meristems that are derived from the *promeristem,* i.e. from the zone of permanent and temporary initials of the root apex.

In some species the initials of the various tissue systems are already discrete immediately adjacent to the central cells, i.e. *closed type.* These initials either represent those of the vascular cylinder, the cortex and the common initials of the protoderm and root cap, e.g. as in *Brassica,* or they represent common initials of the cortex and protoderm and separate initials of the vascular cylinder and of the root cap, e.g. as in *Zea* and *Triticum.* The special initials of the root cap were termed *calyptrogen by* Janczewski. In other species the meristems of the different tissue systems finally become distinct only some distance away from the central cells, i.e. *open type.* In this type common initials for the cortex meristem, root cap, and protoderm, or for the meristems of all the tissue systems (e.g. *Allium*), appear on the periphery of the central cells.

The importance of the above types is queried as both types may occur in the same species but in roots of different developmental stages. Recent research on the root apex has shown that a group of cells at the pole of the stele and cortex have a very low mitotic activity. This group of cells, which appears in the shape of a hemisphere or a disc, was termed the *quiescent centre*. It was found that the apex of a primary root of *Zea mays*, 1 mm in diameter, contains 110,000 cells, which are in one of the phases of mitotic circles, and 600 cells in the quiescent centre. The mitotically active cells laying over the surface of the quiescent centre are regarded as temporary initials.

Existence of the quiescent centre was revealed by various physiological and biochemical techniques, including irradiation and feeding of roots with labelled compounds involved in DNA synthesis. Experiments of *in vitro* culture of isolated quiescent centre showed that it retains the capacity to regenerate a new root, independent of influences derived from other portions of the root. Various views have been expressed on possible causes of the appearance of a quiescent centre in roots. One view suggests control by hormones which may be synthesized in the quiescent centre. A substance may stimulate division at low concentration and inhibit at high concentration. Another view is that there is a competition between cells for supplies of nutrients or hormones.

It was also suggested that the root cap controls quiescence in the particular portion of the meristem. One of the regulators of growth in plants is pressure. Pressure exerted by the rapidly dividing neighbouring cells was thus suggested to cause the inactivity in the quiescent centre. The way the files of cells in the root tip run towards the quiescent centre also suggests that the, pressure extended by the expansion of the temporary initial (around the quiescent centre) and their derivatives prevents quiescent centre from expending and thus its cells from dividing.

It is also necessary to mention here the *Korper-Kappe theory* which was put forward by Schuepp. This theory, similar to the tunica-corpus theory of the shoot apex, is based on differences in the planes of cell division. According to the Korper-Kappe theory the cells divide in a pattern which was termed T-divisions. In the outer regions of the root apex the Kappe consists of cells in which, after the first horizontal division, the lower daughter cell divides longitudinally, i.e. at right angles to the plane of the first division. Thus the planes of the two divisions form a T in a median longitudinal section of the root.

In the Korper—the inner region of the apex—the T is inverted, i.e. the second division takes place in the upper daughter cell. In some families, such as the Gramineae, the position of the boundary between these two regions was found to be constant in relation to the meristems of the various tissue systems, but in other plants, e.g. *Fagus*, the boundary may occur in different positions in the root tip, such as in the middle of the cortex, between the cortex and the epidermis, etc.

Intercalary meristems

Elongation of the shoot axis may be caused by uniform meristematic activity in the internodes or it may occur as a wave of growth extending progressively from the base of the internode in an acropetal direction, as, for instance, in *Helianthus annuus*. Fisher and French, who have studied the manner of extension growth of a large number of species belonging to various families of the monocotyledons, distinguished between two types of meristems in developing internodes—*uninterrupted meristems* and *intercalary meristems*.

The uninterrupted meristem is progressively confined to the upper region of the internode and is continuous with the subapical meristematic region of the stem. Cell maturation within the internode is acropetal. The intercalary meristem represents isolated meristematic regions that are disjunct from the subapical meristematic region and the internodes mature basipetally. The nodes are the first to mature. In most plants with intercalary meristems the region with the cells showing the least degree of differentiation is at the base of the internode, but this region may sometimes be found at higher levels of the internode. In more mature stages the intercalary meristems are separated from each other by fully matured tissues. Finally, they undergo complete differentiation and so disappear.

In the intercalary meristematic region vascular elements differentiate, but they become destroyed by stretching caused by the elongation of the ground tissue. The function of the destroyed tracheary elements is taken over by newly differentiating elements, as has been observed in *Eleocharis acuta*. Destroyed sieve elements are also replaced by newly differentiated ones. In the intercalary meristem of leaves of *Lolium perenne* and of shoots of *Triticum aestivum* total destruction of sieve elements without immediate replacement by new ones was reported. This produces a bottleneck in the transport, which in turn may enhance unloading of the sucrose into the meristematic sink.

The best-known examples of intercalary meristems are those in the stems of grasses, some other monocotyledons, some species of the Caryophyllaceae, Polygonaceae, and articulated species of the Chenopodiaceae and *Equisetum.* Intercalary meristems also occur in the peduncles of the inflorescences of certain plants, in the leaves of many monocotyledons Pinaceae, and others. The gynophore of *Arachis* (groundnut) also elongates as a result of the activity of an intercalary meristem. Internodal elongation in many grasses is brought about by an intercalary meristem, the cells of which divide to form parallel series of cells and which is, therefore, termed rib meristem.

Enlargement of the derivatives of this meristem also contributes to the elongation of the internode. Miltenyi states that the intercalary meristems of grass internodes have no fixed position, but that their position is altered as the internode elongates. At first the intercalary meristematic activity occurs throughout the internode, as has been mentioned above; but after the development of the lacunae, that are present in most grasses, this activity becomes restricted to the peripheral ground tissue in the proximity of and above the nodal plates, i.e. in the joint regions. The meristematic activity of the joints can be reactivated even in mature sterns. Horizontally placed grass shoots show a rapid, negative geotropic reaction. This response is based on cell elongation on the lower side of the leaf-sheath base, which is called the *pulvinus*. The pulvinus of grasses has a radial symmetry and is made up of epidermal, vascular, parenchymatous, and collenchymatous tissues.

In the region of the intercalary meristem of the single elongated internode of *Cyperus alternifolius* cell divisions in the epidermis were found to occur further from the base than in the ground parenchyma. In this species, as well as in *Avena sativa,* there is a gradual decrease in the stomata formed, towards the base of the internode. In those parts of plants that have already undergone a certain degree of differentiation, such as in flowers, fruits, leaves, and stems without special intercalary meristems, the cells continue to divide for a long time after they have been derived from the apical meristem. This type of growth can also be considered as intercalary, but the growth regions are less well defined.

In the last half century many investigators have been attracted to research on meristems, and especially on apical meristems. This interest is due to the great importance of the apex in which the pattern of initiation of leaves, buds, flowers, and the various tissues is determined. The continuity of the tissue systems of the shoot and root results from the activity of the apices. Because the systems of essential elements, such as the tracheary elements, cannot undergo further changes once they are fully differentiated, and because they must be stable and continuous in order to function, the structure of the plant must be determined in the apical meristem.

In long-lived plants with secondary thickening, the continuity of the conducting systems between the young parts of the root and shoot is brought about by secondary vascular tissues which are produced by the vascular cambium, a lateral meristem. The problem of the existence of initials has been widely discussed. The main factor determining which of the cells of the meristem will be a permanent initial is the location of the cell. When a permanent initial in the shoot apical meristem divides, one of the daughter cells, which is located in the proper place, will inherit its function, and the other daughter cell will continue to divide. The descendants of the latter cell will continue to divide until they are in a maturing region, where these cells will undergo differentiation and maturation.

When a permanent initial becomes aborted or injured, another cell, which is in the proper position as related to the other cells in the tissue, will acquire the function of a permanent initial. In the root apex the cells on the periphery of the quiescent centre are regarded as temporary initials. The fact that cells of the quiescent centre do not divide frequently may not exclude the view that they act as initials. They may be regarded as permanent initials which contribute cells at relatively long intervals to the periphery of the quiescent centre where they function as temporary initials.

Chapter 2

The Cell

The cell is the basic component of every organisms. In plants, on the basis of internal organization cells are divided in to *prokaryotic cells* and *eukaryotic cells.* The eukaryotic cells, characteristic of all animals and plants, excluding blue—green algae and bacteria, have a membrane-bound nucleus and other organelles. Prokaryotic cells lack these compartments. With the discovery of the electron microscope and the great progress in biochemical studies, a lot of knowledge has been accumulated on the structure and biology of the cell.

There is a number of books dealing specifically with the ultrastructure and function of plant cells. The protoplast consists of membranous and non-membranous components. When suitably fixed the cross-sectioned membranes are seen under the electron microscope as two dark lines, each about 2.5 nm thick, separated by a lightly stained line, about 3.5 nm thick. This type of membrane is termed a unit *membrane.* It was generally considered that the unit membrane consists of a bimolecular lipid layer covered on each side with a layer of protein.

Today the fluid mosaic model of functional membrane is widely accepted. The cell usually contains a single nucleus, but in some cells, such as the sieve elements of the phloem which are adapted for translocation, the nucleus is generally absent from the mature cell. However, there are also cells which have numerous nuclei, e.g. in some laticifers. A multi-nucleated cell can comprise an entire organism as in some fungi and algae, or multi-nucleated cells may be a transitory stage in the development of a tissue, e.g. in the endosperm of many plants and sometimes in fibres. The accepted view in many cases is that each nucleus together with the protoplasm surrounding it forms a wall-less cell so that the entire multi-nucleate body comprises a group of protoplasmic units. Such a structure is called a *coenocyte.* The coenocyte aroused much interest in phylogenetic and ontogenetic studies.

Two theories exist which deal with the relation of the entire organism and the single cell. According to the cell theory, which was developed about the middle of the nineteenth century, the organism consists, both phylogenetically and ontogenetically, of a complex of an enormous number of cells each of which plays a role in determining the nature of the organism. According to this theory the organism as a unit, to a large extent, determines the nature of the cells.

These two theories are important and for the following reasons, attention was paid to both of them in histological and cytological research of plants. Many aspects of ontogeny, such as the processes of cell division, the origin of vessels and articulated laticifers, the development of idioblasts, etc., were investigated in the light of the cell theory. However, the specialization of the different cells and tissues in the plant and the sites of appearance of the various types of cells and tissues can be explained only on the basis of the organismal theory which regards the organism as a unit.

Cytoplasm

Various minute structures were discovered within the cytoplasm, first with the aid of the light microscope and later with the aid of the electron microscope. These discoveries complicated the exact use of the term *cytoplasm.*

It is now conventional to distinguish between the term *cytoplasm* in its wide sense, i.e. the whole protoplasmic substance surrounding the nucleus, and the terms *groundplasm, ground cytoplasm, cytoplasmic matrix,* or *hyaloplasm* for the portion in which a structure is not clearly established at present.

The plasmalemma frequently exhibits regions which are folded to form an elaborate system of evaginations in tubular form. Such evaginations, together with portions of cytoplasm which they contain, may bud off from the surface of the protoplast.

In electron microscopical sections these regions are often seen as pockets of vesicles and tubules between the plasmalemma and cell wall, and are often referred to as *multivesicular* structures, *plasmalemmasomes,* or *paramural* bodies. It was suggested that these structures may be involved in secretory processes including deposition of wall matrix polysaccharides.

Endoplasmic reticulum (ER)

The endoplasmic reticulum (ER) is one of the membranous structures occurring in the cytoplasm. This is a complex system which consists of two unit-membranes enclosing a narrow space between them. The ER appears in the form of extended cisternae, tubules, or as fenestrated sheets. It may be connected with the nuclear envelope. In a special form the ER occurs in the cytoplasmic strands, *plasmodesmata* traversing the walls of neighbouring cells.

The portion of the ER situated in the centre of the plasmodesma, in tubular form, is called *desmotubule*. The association with ribosomes is interpreted as evidence that the rough ER is involved in protein synthesis. It has been suggested that the ER serves as an intracellular transport system and in some cam has a role in wall formation and in secretion (e.g. of nectar). The ER cisternae may become dilated and accumulate protein and other products. There are suggestions that by dilation ER cisternae or ER vesicles form vacuoles.

Golgi apparatus

Golgi apparatus consists of a system of stacks of flat circular cisternae, each bound by a smooth unit membrane. Each stack is termed Golgi body or *dictyosome*. The margins of the cisternae may be simple in shape or fenestrated. Near the dilated edges of the cisternae there is usually a number of vesicles positioned in such a way as to suggest that they have budded off from the cisternal margins. In some Golgi bodies rod-shaped elements in parallel arrangement have been seen in the very narrow spaces between the individual cisternae. It has been suggested that these elements might link the cisternae.

The Golgi bodies are concerned with secretion processes and have a polar structure. In active bodies, in addition to the production of many vesicles, distal cisternae (on the maturing or secreting face) may break up into vesicles. New cisternae are formed on the opposite face (proximal face or forming face) of the Golgi body. The new cisternae appear to be produced from membrane components derived from the ER.

The Golgi bodies are mainly involved in the secretion of sugar (in nectar secretion), polysaccharides (cell-wall material, mucilage), and polysaccharide protein complexes (certain mucilages).

Mitochondria

Mitochondria are organelles which can be seen with the light microscope when living cells are stained with Janus Green B. In electron-microscopical sections they appear in various forms: spherical, elongated and sometimes lobed. They are about 0.5 μm in diameter and up to 6 μm in length and are bound by an envelope consisting of two unit-membranes.

The inner membrane forms invaginations (*cristae*) into the matrix. The matrix is mainly made of proteins. The mitochondria contain ribosomes which are smaller than those of the cytoplasm and DNA fibrils, but their genetic capability is limited.

The mitochondria are concerned with processes of energy conversion and contain many enzymes, especially those of the Krebs cycle. Glandular cells, which are highly active, contain numerous mitochondria. The formation of new mitochondria is most likely by division.

Plastids

Plastids are special organelles of plant cells. They vary in form, size, and pigmentation. The principal types of plastids are *chloroplasts*, *chromoplasts*, and *leucoplasts*. Chloroplasts arc green as a result of the pigment chlorophyll which predominates in them. Chromoplasts are usually yellow, orange, or red because of the carotenoid pigments. They occur in petals, in some ripe fruits, and in some roots (e.g. carrot roots). Leucoplasts are non-pigmented plastids, usually located in tissues not exposed to light, and they store plant products such as starch (*amyloplasts*), proteins (*proteinoplasts*), and fats (*elaioplasts*). Starch, phytoferritin (an iron-protein complex) and lipids in the form of globules (plastoglobuli) may be present in various plastids, including chloroplasts. Leucoplasts of tissues which become exposed to light may develop into chloroplasts (e.g. in the potato tuber). The proteinoplasts of *Helleborus corsicus* were observed to initiate as amyloplasts; later the starch disappears and the main product of the plastid becomes protein.

All types of plastids are derived from minute bodies, *proplastids* which are present in egg cells and meristematic cells. Proplastids and plastids in young cells multiply by division. The plastids are bound by an envelope consisting of two unit membranes. Internally the plastids contain a membrane system embedded in a proteinaceous matrix or *stroma.* The stroma contains fibrils of DNA and ribosomes, which are smaller than those of the cytoplasm. The membrane system is in the form of flattened sacs called *thylakoids.* The thylakoid system is developed to various degrees in the different types of plastids. It is least developed in proplastids where only few or no thylakoids are present.

With the differentiation of plastids flattened vesicles bud off from the inner membrane of the envelope, proliferate, and form thylakoids which align themselves into characteristic systems. Each granum is composed of a series of disc-like thylakoids stacked one upon the other like a pile of coins. The grana are interconnected by the frets that traverse the stroma. It is now accepted that these two types of thylakoids are interconnected in such a way that the spaces within them are continuous.

In addition to the light harvesting system the chloroplasts contain the enzymes responsible for the fixation of carbon dioxide into sugar. In plants grown in the dark the developing leaves and stems lack the green colour. They become *etiolated.* In such a case the vesicles derived from the inner membrane of the plastid envelope develop into a paracrystalline lattice called *prolamellar body.* Plastids of this kind are called *etioplasts.* When a darkgrown plant is placed in light the prolamellar body gives rise to the thylakoid system characteristic to chloroplasts. *Chromoplasts* vary greatly in shape and size and are often derived from chloroplasts (e.g. in many fruits), but may also develop directly from proplastids. The differentiation of chromoplasts involves synthesis of carotenoid pigments. These pigments may be stored in the lipid globules (e.g. in petals of *Ranunculus repens* and yellow fruits of *Capsicum* and *Citrus),* or in protein fibrils (e.g. in red fruits of *Capsicum).* In the carrot root the caroten is in a crystalloid form surrounded by lipoprotein envelope. In tomato fruits two kinds of pigment bodies were found-long, narrow crystaloids and round globules. Chromoplasts in many plant organs are capable of reverse differentiation into chloroplasts.

Both plastids and mitochondria contain DNA and ribosomes. Thus the two organelles are potentially autonomous. The indication of autonomy has led some researchers to the hypothesis that plastids and mitochondria originated, in the course of evolution, from primitive prokaryotes, which were enclosed in primitive eukaryotic cells and became there established as permanent symbiotic bodies.

Microbodies

Small bodies, 0.5-1.5 μm in diameter, which occur in the cytoplasm of a variety of tissues are called microbodies. They are bound by a single membrane and their matrix appears granular or fibrillar. They contain enzymes that vary in accordance to the type of cell or tissue in which they are present. In leaves of higher plants, closely associated with chloroplasts, microbodies called *peroxisomes* are found.

These microbodies are sites for oxidation of glycollic acids, a product of carbon dioxide fixation. The oxidation of this compound results in a release of carbon dioxide and oxygen. Microbodies which occur during the germination of those seeds which store fats as a reserve, and contain the enzymes necessary for the breakdown of fatty acids to acetyl-CoA and the synthesis of succinate from acetyl CoA, are called *glyoxysomes.*

Ribosomes

Ribosomes are small particles (17–20 nm in diameter) that occur free in the cytoplasm, on the outside of the membranes of the endoplasmic reticulum, and in the nucleus, chloroplasts, and mitochondria. They consist of RNA and protein, mainly histone.

Ribosomes united into clusters *(polyribosomes* or *polysomes)* are involved in protein synthesis, i.e. in the assembly of polypeptide chains from amino acids. The information specifying the amino-acid sequence of the polypeptide chain is brought from the nucleus by the messenger RNA molecules. When the messenger RNA molecules associate with the ribosomes, the information it contains is translated by transfer RNA molecules located in the cytoplasm.

Spherosomes

Spherosomes are spherical lipid bodies which become electron opaque after fixation with osmium tetroxide. There is no unanimous interpretation of the nature and structure of these bodies. In the view of some investigators the spherosomes are surrounded by a membrane, whereas in the view of others a surface skin consisting of an outer layer of oriented lipid molecules is formed in response to the aqueous cytoplasm surrounding them. It has been suggested that spherosomes originate as oil-containing vesicles detached from the ER.

Microtubules

Straight, elongated, hollow structures composed of globular protein subunits are called microtubules. Their average diameter is 23—27 nm. Microtubules occur in the peripheral cytoplasm close to cell walls still growing in area and thickness, in the mitotic and meiotic spindles, and in the phragmoplast that arises between the daughter nuclei at the telophase.

Similarity in the alignment of microtubules in the cytoplasm close to the plasmalemma and the cellulose microfibrils in the cell wall, led to the view that the microtubules might direct the orientation of the developing microfibrils.

Nucleus

Nucleus is usually more or less spherical, though nuclei with other shapes have also been observed. The nucleus is surrounded by an envelope *(nuclear envelope)* and contains the nuclear matrix *(nucleoplasm)* and one or more *nucleoli.*

In the nucleoplasm chromosomes consisting of deoxyribonucleic acid (DNA) and proteins are present. The complex of DNA and protein in the chromosomes which has an affinity to basic dyes is called *chromatin.* In the interphase (between nuclear divisions) the chromosomes are uncoiled and cannot be distinguished. At this phase condensed, deeply staining chromatin masses are often seen in the nucleoplasm. This chromatin is called *heterochromatin.* The chromatin which stains less deeply than heterochromatin is called *euchromatin.*

During the interphase DNA synthesis takes place in preparation for the process of replication of chromosomes, The nuclear envelope consists of two unit-membranes. The narrow space between these membranes is called *perinuclear space.* Connections exist between the nuclear envelope and the ER. The nuclear envelope has pores which are usually distributed in a more or less regular pattern.

The nucleoli are very dense, granular, and fibrillar in structure, and are not bound by a membrane. They are often seen associated with some chromatin. They contain RNA. DNA, and proteins. Light-stained regions, which are often seen in the nucleoli, are commonly referred to as "vacuoles".The nucleus carries the information for the cell protein in its DNA. The special kind of RNA, the messenger RNA, is transcribe on the DNA. The latter is transported to the cytoplasmic ribosomes where the synthesis of proteins, mainly enzymes, occurs.

According to the genes present in the chromosomes the particular kinds of messenger RNA are synthesized. Differentiation can occur without differences in the DNA, as different cell types undergo different courses of development, according to which genes arc active or depressed.

Vacuoles

A vacuole is a watery cell compartment surrounded by a membrane, the *tonoplast.* They occupy 90% of the volume of most plant cells. It contains a variety of organic and inorganic substances, such as sugars, proteins, organic acids, phosphatides, tannins, flavonoid pigments, and calcium oxalate. Some substances in the vacuole may occur in solid form and may even be crystalline. Various views exist as to the initiation of the vacuoles:

i) from pre-existing vacuoles which multiply by fission, and after cell division each daughter cell receives a number of vacuoles;

i) by a *de novo* process, by attraction of water to a certain localized region in the cytoplasm and the formation of membrane around it;

iii) From Golgi vesicles;

iv) Dy dilation of ER cisternae or vesicles derived from the ER.

In any case it seems that there is more than one way by which vacuoles originate. The vacuoles function in regulation of the water and solute content of the cell, i.e. in osmo-regulation, in storage, and in digestion. There is evidence that vacuoles contain digestive enzymes capable of breaking down of cytoplasmic components and metabolites.

The hydrolitic activity of vacuoles resembles that of lysosomes of animal cells. The origin of the digestive enzymes of the vacuoles may be in the ER and the Golgi apparatus from where they are transported to the vacuole by membrane-bound vesicles. The amount of enzymes can change during the life of the cell and may be produced to different extents in different cells.

Proteins

Protein, in the form of cuboidal crystalloids is found in the cells of the peripheral parenchyma of the potato tuber and in the fruit parenchyma of *Capsicum.* Crystalline and amorphous protein are found together in aleurone grains in the endosperms and embryos of many seeds. The development of *aleurone grains* in the seed of *Ricinus* has been described in detail by Frey-Wyssling. These grains are formed from readily soluble proteins with globular molecules and relatively low molecular weight which accumulate in vacuoles of the storage cells where they crystallize. From these liquid vacuoles water is lost by dehydration.

This causes the various vacuolar components to precipitate according to their solubility. In *Ricinus* the first substance to be precipitated is the almost insoluble phytin: this substance forms the globoid. Next, the corpuscularly dispersed reserve proteins precipitate out in a lattice to fill the remaining space of the vacuole and so form the crystalloid part of the aleurone grain. Finally, the remaining liquid, which now contains soluble albumin, solidifies to form a homogeneous substance which surrounds both the globoid and crystalloid. On mobilization, these reserve substances are digested in the reverse order.

Crystals and silica bodies

Many plants are known to deposit in their cells excess Inorganic materials consisting mostly of calcium salts and silicon dioxide. The first occur as crystals and the second as silica bodies. Crystals of various forms are often present in plant cells.

The most common crystals are those of calcium oxalate, which appear in different forms. Crystals can be found in cells resembling those neighbouring them but lacking crystals, or they may be confined to special crystal-containing cells, i.e. *idioblasts.* In some plants, e.g. in *Zebrina,* the idioblasts containing raphides form long files. Crystals are formed within vacuoles and are surrounded by an envelope.

It has been suggested that in mature idioblasts, the envelope is suberinic in nature. Calcium carbonate crystals are rare in higher plants. This type of crystals is associated in some plants with ingrowths of the cell wall known as cystoliths. The appearance, location, and type of crystals is often used in taxonomic classification.

Silica bodies

Silica bodies occur in the epidermis of plants such as Gramineae, Cyperaceae, Palmae, Rapataceae, and in ray cells and axial parenchyma of the wood of some dicotyledons, e.g. Lauraceae.

Tannins

Tannins can be found in the different parts of the plant, especially in leaves, in periderm, in vascular tissue, in unripe fruits, in seed coats, and in tissues of pathogenic growths. The tannins are a heterogeneous group of phenol derivatives. In microscopical sections they usually appear as granular masses or bodies coloured yellow, red, or brown.

Tannin-containing cells may be interconnected or tannins may be found in isolated specialized cells (idioblasts). Within the cell the tannins may be found in the vacuole or in the form of droplets in the cytoplasm, and sometimes they penetrate into the cell wall, as, for instance, in cork tissue. Tannins are thought to protect the plant against dehydration, rotting, and damage by animals. Tannins are used commercially, especially in the industry of the tanning of animal skins to obtain leather.

Plant pigments

The pigments are usually found in the plastids and in the vacuole. The green colour is due to *chlorophyll* which is found in the chloroplasts. In the same plastids *carotenoids,* the yellow to red pigments, are also found but they are masked by the chlorophyll.

The carotenoids become noticeable when there is little or no chlorophyll as is the case in chromoplasts. Sometimes the visible colour of petals is the result of a few pigments occurring in the same cells. For instance, chloro- or chromoplasts can be found together with anthocyanins.

White petals may be devoid of pigments, and the colour seen results from the reflection of light from the petals which are opaque due to the presence of numerous large intercellular spaces that are filled with air. The colouring of autumn leaves is the result of various processes as well as the combination of different pigments.

With the gradual death of the leaf the chlorophyll breaks down into colourless substances and the carotenoids become visible, making the leaf appear yellow. The red and purple colours are from oxidation products of flavonoids. These colours are most brilliant when formed in the presence of sugars in leaves exposed to strong light. Autumn colours, which result from the combination of small amounts of chlorophyll and carotenoids and greater amounts of anthocyanins together with tannins and various uncommon pigments and the browning of the cell wall, are best developed in the cold temperate zones.

Shape and arrangement of cells

Cells may be ellipsoidal, ovate, cylindrical, flattened, prism-like, star-shaped, fibre-like, and lobed. Parenchyma cells are Usually from 10 to 100 p in diameter, but in fleshy fruits and the pith of stems larger cells can be found. Fibres are usually about 1-8 mm long but fibres 55 cm long are also known, e.g. in *Boehmeria*. As the cell volume increases in the meristematic region the primary elastic cell wall tends to assume the smallest possible surface area, i.e. the form of a sphere.

After mitosis, forces act that tend to impart to the cell a spherical form but as there are no intercellular spaces in the meristem, the cells, which are densely arranged, become polyhedral in shape. The basic shape of cells is a 14-faced polyhedron. However, in plant tissues cells with 12, 13, 15, 16, or more faces are found. Most of the faces of the cell wall are, according to Matzke, pentagonal, but tetragonal and hexagonal faces can also be found. Similar structure was found in bubbles of soap foam, and also in experiments where lead shot was subjected to sufficient pressure to cause the elimination of the air spaces. As a result of the continued increase of cell volume during growth, the number of wall faces increases above 14. This makes it impossible for all the sides to remain in contact with all the sides of the neighbouring cells and so intercellular spaces develop. In some tissues the intercellular spaces reach relatively large dimensions and then they are referred to as air spaces, ducts, etc. Such spaces can develop in two ways:

i) By the separation of neighbouring cell walls, as in the development of the resin ducts in *Pinus;* this type of development is known as *schizogenous* development;

ii) By the disintegration of the cells in the place where the space develops, as in the essential on cavities in the peel of citrus fruits, this type of development is known as *lysigenous* development

In some cases spaces are formed by these two methods together and then the development is known as *schizolysigenous* development. The intercellular spaces in the protoxylem are sometimes formed in this way. The intercellular spaces can be irregular and variable in shape or they may form a distinct and permanent system as in many water plants, in the banana leaf and other plants. All cells develop from existing cells by cell division. The young cells in the growing regions are all relatively small, as they mature their size and shape alter in accordance with their physiological function. As the cell wall is already present at the earliest stages of development of the cell it takes part, together with the protoplast, in the processes of cell growth.

When a group of cells grows together uniformly, the cells of the group take up different positions and shapes, but the relationships between the neighbouring cell walls do not change and no new areas of contact are formed between the cells: this type of growth is known as *symplastic growth.* In many cases, pit-free tips may grow individually along walls of neighbouring cells and form new areas of contact. In this type of growth, termed *intrusive growth,* a temporary loosening of the middle lamella between, the cells in the growing region must occur. Intrusive growth takes place in many elongating fibres, in branches of some sclereids, in non-articulated laticifers, and in the daughter cells of a fusiform initial of non-storied cambium which has divided anticlinally.

The cell wall

The cell wall was discovered in the seventeenth century before the presence of the protoplast was recognized, and since then many researchers have investigated it. Various methods-chemical, physical, and morphological—have been used. These investigations were assisted by advances in organic chemistry, X-rays, the use of the light and polarizing microscopes, and, recently, by the electron microscope. The cell wall grows when in contact with the protoplast but outside of it. In pollen grains only it has been suggested that the outermost part of the wall, the exine, may develop by contributions from both sides of the young pollen wall, i.e. from the protoplast within the pollen grain and from the tapetum which surrounds the developing pollen grains.

On the basis of the development and structure of plant tissues it is possible to distinguish the following principal three layers in the cell wall:

i) the *middle lamella* or the *intercellular layer* or *substance;*

ii) the *primary wall,* and

iii) the *secondary wall.*

The *middle lamella* is the cement that holds the individual cells together to form the tissues and accordingly, it is found between the primary cell walls of neighbouring cells. It is an amorphous substance, inactive in term of polarized light (isotropic). In supporting tissues it may also fill the intercellular spaces. The middle lamella consists mainly of pectic substances. The enzyme pectinase and chemical reagents which solubilize pectins disintegrate tissues into their individual cells. This procedure is called *maceration.*

The *primary, wall* (P-wall) is the first true cell wall which develops on the new cell. In many cells it is the only cell wall, as the middle lamella is regarded as intercellular substance and not a wall proper. The primary wall is that part of the cell wall that develops in cells or portions of them which are still growing. This wall is optically active (anisotropic).

The *secondary wall* (S-wall) is formed on the inner surface of the primary wall. It begins to develop in cells, or parts of them, that have ceased to grow. The secondary, wall is very strongly anisotropic, and layering can be observed in it. In the majority of tracheids and fibres three layers-the *outer layer* (S_1), the *central layer* (S_2), and the *inner layer* (S_3)—can be discerned in the secondary wall.

Of these layers the central layer is usually the thickest. In some cells, however, the number of layers may be more than three. Some authors use the term *tertiary wall* for the inner layer of the secondary wall. It should be mentioned that some investigators use the term compound *middle lamella* when dealing with wood tissue. This term is used to refer to the complexes of lignified layers which appear more or less homogeneous when examined, without pretreatment, under the light microscope. The compound middle lamella may be three-layered when it refers to the middle lamella proper and the adjoining primary walls, or five-layered when it refers to the middle lamella proper, the primary walls, and the outer layer of the secondary walls of the adjoining cells.

When fibres and tracheids are examined without special treatment under the ordinary light microscope, layers are visible in cross-section. With special treatment and with the high-power magnification of the light microscope, finer lamellation can be seen. The coarser layers represent groups of lamellae. The lamellae can be concentric, radial, or have a complicated arrangement. When the cell wall is allowed to swell under the influence of suitable reagents, fine structures can be observed within the lamellae themselves.

By such methods Bailey and others found that the cell wall is built of a system of microscopic threads-the fibrils. It has been revealed that the wall consists of two continuous interpenetrating systems, one of which is the cellulose fibrils and the other the continuous system of microcapillary spaces. These spaces may be filled with lignin, cutin, suberin, hemicelluloses, and other organic substances, and even mineral crystals and in fresh tissue aqueous solutions. The material between the fibrils forms the non-cellulosic matrix.

In tracheary elements and sclerenchyma cells the layers seen with the ordinary light microscope are usually the result of the different quantities of lignin, pectic substances, hemicellulose, or other organic substances deposited in the interfibrillar spaces of the cellulose, or they may be due to the presence, in certain cells, of layers poor in cellulose, or of the different orientation of the microfibrils in the various wall layers.The lamellation seen in the secondary wall is often the result of the different density of the fibrils. In the denser, darker areas, the fibrils are more numerous per unit area and they are more tightly packed. In the less dense, lighter areas, the fibrils are looser and the capillary spaces between the fibrils are larger. Most fibres have lignified walls.

Electron microscope studies of such fibres have confirmed previous assumptions that each lamella is composed of two parts, one consisting mainly of cellulose and one predominantly of lignin. The lignin permeates the cellulosic part formed previously, impregnates it, and cements it to the preceding lamella. In heavily lignified walls it is possible to dissolve the cellulose and retain the lignin only, or the lignin alone can be dissolved and the cellulose retained. In this way the component retained gives, as it were, a negative image of the component which has been dissolved. This phenomenon not only proves that the lignin is found in the elongated interfibrillar spaces of the cellulose but also that these capillary spaces are continuous.

In order to demonstrate the presence of the two parallel, three-dimensional systems, the cellulosic fibrillar network and the network of interfibrillar microcapillary spaces, Bailey also used methods other than those described above. For example, he succeeded in crystallizing iodine in the elongated, microcapillary spaces, thus demonstrating their presence and the orientation of the fibrils in the different layers of the wall.

Further clarification of the fine structure of the cell wall is based on the use of the electron microscope. The photographs, which were made with the electron microscope, revealed the fine *microfibrils* which cannot be seen by means of the ordinary light microscope. The results of research with the electron microscope have in general confirmed the theories of Bailey on the structure of the wall. The morphological structure of the cellulose in the cell wall, as is known today, can be summarized in the following way.

Within the cell wall differently built lamellae are recognized, each of which consists of fibrils. By using certain techniques it is possible to distinguish macrofibrils with the aid of the light microscope. These form a three-dimensional network. The network is interwoven with a parallel network of microcapillary spaces occupied by non-cellulose substance. The chain molecules are arranged in bundles which were termed *micellae*.

The hypothesis of the presence of micellae was proposed by Nageli in the last century. According to him the micellae are the individual units arranged in a permanent order within an intermicellar matrix. With the aid of the polarizing microscope the crystal-like nature of the micellae was proven. From the results of various investigations, especially those made with X-rays, investigators came to the conclusion that the micellae consist of parallel chains of glucose residues which have characteristic and permanent distances between them.

As a result of extensive research carried out by botanists, chemists, and physicists, several theories were suggested which attempted to explain the organization of the cellulose molecules in the cell wall. The chain-like cellulose molecules are regularly arranged in bundles. Each such bundle, which forms an elementary fibril, consists of about 40 cellulose molecules and is about 3.5 nm wide and 3 nm thick.

The elementary fibril is in greatest part crystalline. Only very small parts of it, which are presumably arranged at random, may be paracrystalline. The number of glucose residues in cellulose molecules of fibre cells was found to vary from 500 to 10,000 and the length of these molecules varies from 0.25 to 5 μm. Most of the above is based on the results of research made on the secondary cell wall, but recently much attention has been paid to the structure of the primary wall. The primary wall is similar in structure to the secondary wall in that it consists of anisotropic (crystalline) cellulose microfibrils and a non-cellulotic matrix.

The interfibrillar matrix usually contains pectic compounds and hemicellulose. The outstanding mechanical property of the cellulose is tensile strength, while under compressive stress cellulose fibrils bend. This effect is prevented in supporting cells by replacement of the plastic interfibrillar matrix in the cell wall by solid substances which harden it. This introduction into the cell wall of additional substances in place of the matrix in which the cellulose framework is embedded is termed *incrustation.* The most important incrusting process in higher plants is *lignification,* but in many cells, substances such as suberin, cutin, waxes, quinones, tannins, and other organic and mineral substances may incrustate the walls.

The primary wall of many cells has a lamellate structure. The angle at which the microfibrils cross in the primary wall differs in different cells as well as in the various lamellae and in different parts of the wall of a single cell. The classical debate as to whether the growth of the cell wall is accomplished by *intussusception* or by *apposition is* still continued. According to the first opinion the material of the new wall is laid down between particles of the existing substance of the expanding wall. According to the second opinion the growth is due to the centripetal addition of new layers one upon the other.

New theories, based on electron microscopic research, have been developed as to the manner in which the primary wall grows. Frey-Wyssling and Stecher, for instance, suggested that the primary cell wall grows in a way that has been termed *mosaic growth.* According to this view the fibrillar texture in certain wall areas becomes loosened as a result of turgor pressure and afterwards mended by deposition of new microfibrils in the gaps caused by the stain. The loosening of the fibrillar network requires the plasticizing of the wall matrix in the expanding areas. Growth hormones available to the cell and proteins and enzymes occurring within the cell wall are involved in regulating the process of extension growth.

Another concept of growth is the theory of *multinet growth*. According to this theory the thickening and increase in surface area of the primary wall is brought about, in many cases, by the separation of the crossed microfibrils and alteration in their orientation, in the earliest formed lamellae, from being almost transverse to almost longitudinal. New lamellae with denser, crossed, and almost transversely orientated microfibrils are added centripetally.

Regarding the cellulose microfibrillar system it can be assumed that according to the multinet growth theory the fibrils are added to the growing cell wall by apposition, whereas in view of the mosaic growth concept intussusception also takes place. It has, of course, to be kept in mind that the wall matrix is continuously secreted not only into the lamellae adjacent to the cytoplasm but also into the outer lamellae.

The wall in a cross-section appears to be built of layers, because of the different composition and structure of the lamellae which are added continuously as the wall grows. The difference in the structure is brought about, as explained above, by the differences in density and orientation of the cellulose microfibrils, the presence of different quantities of lignin, etc. These features result in differences in the refraction of light so that the layers of the wall are emphasized. Among the physical properties of the cell wall are tensility, strength, resistance to compression swelling, and permeability properties.

Because of the large amount of cellulose in the wall, the properties of the wall are mainly determined by those of cellulose; the other substances add some characteristics, or alter slightly, those of the cellulose.

One of the important properties of cellulose is its ability to withstand stretching because of its elasticity. The lignin increases the resistance of the wall to pressure and so prevents the folding of the cellulose microfibrils. The orientation of the microfibrils in the different lamellae of the wall no doubt is an important factor in determining the strength of the wall.

Pits on cell wall

The primary walls of young cells stretch and increase in surface area and thickness as the cell grows. However, certain areas of the wall generally, remain thin: these portions are termed primary *pit fields* occasionally also called *primary pits or primordial pits*. Some times, when the primary pit fields are numerous and deeply sunken, the wall in which they *occur* appears beaded in cross-section.

A characteristic feature of the primary pits of living cells is the presence of concentrations of very thin protoplasmic strands, i.e. the *plasmodesmata*. The plasmodesmata connect the protoplasts of neighbouring cells. They are present almost in all living cells of higher plants, the living protoplasts of which are, therefore, united to form a single unit. As a result of this the plant body can be divided into two major compartments: The interconnected protoplasts, called *symplast*, and the compartment external to the protoplasts, the *apoplast*.

The latter consists of the cell walls, intercellular spaces, and the lumen of non-living cells, such as tracheary elements and most fibres. The diameter of the plasmodesmata is at the limit of resolution of the light microscope, and they can be observed only by suitable staining or by impregnation with heavy metals such as silver or mercury. Plasmodesmata are relatively easily seen in the endosperm of seeds, such as those of *Phoenix*, *Aesculus*, and *Diospyros* and in the cotyledons of' some plants. On plasmolysis the protoplast withdraws usually from the wall in all places except where plasmodesmata are present. Plasmodesmata usually occur in groups, but sometimes they may be more or less evenly distributed over the entire wall. When in groups the plasmodesmata usually occur in the primary pit fields as has been demonstrated in cambial cells. In mature living cells with secondary walls large groups of plasmodesmata traverse the pit membranes.

In the intercellular canals housing the plasmodesmata the plasmalemma of the two neighbouring cells is continuous and that through the centre passes a tubule, called *desmotubule,* which is considered to be a derivative of the ER and which was observed to be continuous with it. The desmotubules may contain an axial *central rod.* The space between the desmotubule mid the inner face of the *plasmalemma* is called *cytoplasmic annulus.* The diameter of the lumen of plasmodesmata is about 30-60 nm. The external diameter of the desmotubule is 16- 20 nm. There is a controversy in the literature as to the possibility that new plasmodesmata can be produced between cells or portions of cell after completion of the cell plate, e.g. during intrusive growth, in graft junctions, between tyloses, between parasitic and host cells, etc.

In a variety of cells branched plasmodesmata have been observed to occur, e.g. between companion cells and sieve elements, in pit membranes of living fibres, e.g. of *Tamarix* and at other sites. According to Krull the branched plasmodesmata of the longitudinal walls of the cortical cells in *Viscum album* develop during wall extension, the branches being secondarily divided plasmodesmata. The branches depart from a median cavity or nodule to the cells on both sides. There is also a possibility that such branches are primary plasmodesmata which became joined by dissolution of a median portion of the common wall of neighbouring cells. Plasmodesmata play an important role in the transport of materials and the relay of stimuli.

Certain portions of the cell wall remain thin even as the secondary wall is formed and they, therefore, consist only of primary wall material. These areas which are of variable shape, are called pits. Some authors use the term pit to refer only to the pit cavity together with the primary wall, which closes the pit. Others use the term pit to refer to the above structures together with that part of the secondary wall that Surrounds the pit cavity. Pits can develop over the primary pit fields and then one or more pits may develop in the pit field, or pits develop on those parts of the primary wall devoid of pit fields. On the other hand, the primary pit fields can become completely covered by the secondary cell wall. The pits are apparently areas through which substances pass from cell to cell.

The concentration of plasmodesmata in living cells in the region of the pit membranes is all additional proof of the pit being a channel of exchange. Generally, each pit has a complementary pit exactly opposite it in the wall of the neighbouring cell. Such pits form a morphological and functional unit called the *pit-pair.* The cavity formed by the break in the secondary wall is called the *pit cavity.* The membrane, built of the primary cell walls and middle lamella that separates the two pit cavities of the pit-pair is called the *pit membrane* or *closing membrane.*

The opening of the pit on the inner side of the cell wall, i.e. on that side facing the lumen of the cell, is called the pit *aperture.* Two principal types of pits are recognized *simple pits* and *bordered pits.* The main characteristic of bordered pits is that the secondary wall develops over the pit cavity to form an overarching roof with a narrow pore in its centre.

In a simple pit no such development of the secondary wall is present. If the two pits of a pair are simple, a *simple pit-pair is* formed: if the two pits are bordered, a *bordered pit-pair,* if one of the pits is simple and the other bordered, a *half-bordered pit-pair*. If the pit has no complementary pit in the adjacent cell or if it is opposite an intercellular space it is termed a *blind pit*.

Sometimes two or more pits are found opposite one large pit such an arrangement is called *unilateral compound pitting*. The pit cavity of a simple pit may have the same diameter over its entire depth, or it may widen or narrow towards the pit aperture. In places where the secondary wall is very thick, the pit cavity has the form of a canal. Sometimes this canal is branched towards the outer layers of the cell wall and then the pit is called a *branched* simple pit. Such pits arise from the fusion of several pits during the centripetal addition of layers to the secondary wall. The bordered pit is more complicated than the simple in its structure and is variously shaped. In the bordered pit that part of the pit cavity that is formed by overarching of the secondary wall is called the *pit chamber* and the opening in the secondary wall that faces the cell lumen is called the pit aperture.

If the secondary wall is very thick a canal the *pit canal*—is formed between the cell lumen and the pit chamber. In the pit canal two openings are distinguished—that facing the cell lumen is termed the *inner aperture* and that nearest the pit chamber, the *outer aperture*. In some plants there are bordered pit-pairs in which the pit membrane is thickened in its central portion: this thickening, which is of a primary nature, is disc-shaped and is termed the *torus*. The diameter of the torus is wider than that of the pit aperture. Barman describes the occurrence of thickenings, other than the torus, on the pit membrane. These thickenings may be radial or tangential in relation to the torus. The torus found in the bordered pits of *Cedrus* is fringed or scalloped on its circumference. This feature is a characteristic of *Cedrus,* and as such it aids in the identification of the wood of this genus. In tracheids of many conifers the pit membrane around the torus—the *margo*—is porous. During differentiation of the margo the cell wall matrix is dissolved so that only the cellulosic microfibrils remain. The pit membrane is usually flexible and, tinder certain conditions, the torus can be pushed against one of the pit apertures.

When the torus is in the median position, i.e. in the middle of a pit-pair, water can easily pass from one tracheid to another. In a pit-pair where the torus I is in a lateral position, i.e. pressed against one of the pit apertures, the passage of water is prevented. Most of the tort in late wood and all of them in the heartwood are always in a lateral position and the flexibility of the pit membrane is lost. In this condition the pit is called *aspirated.*

During differentiation of the bordered pit the cellulose microfibrils of the margo become organized into a system of coarse radially oriented bundles. At the same time circular oriented fibrils form the origin of the torus. The microfibrillar system of the margo and the torus of differentiating tracheids is embedded in amorphous matrix, consisting of hemicelluloses and pectins. During the process of maturation the amorphous embedding substances disappear apparently by enzymatic action and the margo becomes perforated. The presence of a torus is especially characteristic of the bordered pits of the Gnetales, of *Ginkgo,* and most of the Coniferales. Tori occur only rarely in the Ophioglossales and in the angiosperms.

In some dicotyledons thin, simple, or branched sculpturings are present on the secondary wall that forms the pit chamber or around the pit aperture. Such pits are called *vestured pits* and the sculpturings may have various shapes. Because of the special properties of light refraction and of staining, these pits appear as if porous or net-like in surface view and, thus, were once termed sieve-pits. Vestured pits are found in the tracheary elements of the secondary wood of certain dicotyledonous genera and species, such as some of the Leguminosae, Cruciferae, Myrtaceae, and Caprifoliaceae.

Vestured pits are found in phylogenetically more developed xylem and therefore they are considered to be an advanced form of pit. The shape of the pit aperture can be the same as that of the pit chamber or different from it. The pit aperture may be round, elliptic, or linear. As the walls continue to thicken, the pit chamber becomes smaller and the pit canal between the inner and outer apertures becomes longer. In such pits the inner aperture often becomes long and narrow, as seen in surface view, and, in very thick walls, its longitudinal axis may be longer than the diameter of the pit chamber.

When the inner aperture is large and linear, narrow, or elliptic, and when the outer aperture is small and circular, the pit canal has the shape of a flattened funnel. The elongated inner apertures of such a bordered pit-pair may be parallel or crossed. This type of pit occurs mainly in fibre-tracheids. Bordered pits found in tracheary elements vary in shape and arrangement. When the pits arc distinctly elongated or linear and arranged in ladder-like tiers the arrangement is termed *scalarifrom pitting*. When the pits are circular or only slightly elongated and the outlines are elliptic there are two possible ways in which they may be arranged on the wall: in horizontal lines, i.e. *opposite pitting*, or in diagonal lines, i.e. *alternate pitting*. When the pits are crowded the outline of the opposite pits becomes rectangular or square, and that of the alternate pits hexagonal. Tracheary elements have especially well-developed bordered pits in those regions where they are adjacent to other tracheary elements. In the regions of contact with parenchymatous cells reduced bordered pits are sometimes found.

In addition to the pits many other sculpturings exist on the cell walls. These include, for example, the perforations in the end walls of the vessel members, various thickenings on the inner surface of cell walls, such as wall thickenings in the protoxylem elements, spiral thickenings on the inner surfaces of pitted secondary walls, Casparian strips of endodermal cells, thickenings in the walls of the endothecial cells of pollen sacs, and external projections formed partly by the wall itself and partly by deposits, e.g. of cuticle on epidermal cells and of external layers on spores and pollen grains. Here only two structures will be discussed—*crassulae, trabeculae*.

Crassulae are linear or crescent-shaped thickenings of the primary wall and middle lamella which occur between bordered pits or small groups of these pits. The crassulae may sometimes surround the pits. They represent the borders of the primary pit fields of the young cell from which the element developed. Crassulae are well developed in the tracheids of certain gymnosperms. *Trabeculae* are rod-shaped thickenings of the wall which traverse the cell, lumen radially. They usually appear in radial rows in the wood elements.

Chapter 3

Plant Tissues

A tissue can be defined as a group of cells with common origin, structure, and function. When dealing with the tissues of higher plants a more flexible definition is necessary. If we could base our descriptions on elements, i.e. the individual types of cells, it would be easier to define these units. However, difficulties would also arise from such a classification because of the transitional forms present. We also know now that experimental treatments can cause one type of cell to change into another. Today a complex of cells of common origin is generally understood by the term tissue.

The tissues in the plant body are classified on the following bases: according to their position in the plant; the cell types of which they consist; their function; the manner and place of their origin; and their stage of development.

Tissues are also divided into *simple* and *complex tissues* according to the number of cell types that they comprise. A simple tissue is homogeneous and consists of only one type of cell, while a complex tissue is heterogeneous and comprises two or more cell types. *Parenchyma*, *collenchyma*, and *sclerenchyma* are simple tissues, while *xylem*, *phloem*, and *epidermis* are complex tissues. Elements of parenchyma, sclerenchyma, and other types may be included in complex tissues.

Parenchyma

Parenchyma of the primary plant body develops from the ground meristem, and that connected with the vascular elements from the procambium or cambium. The phellogen in many plants also produces parenchyma (the phelloderm).

Parenchyma consists of living cells of differing shape and with differing physiological functions. By the term *parenchyma* we generally refer to tissues which exhibit relatively little specialization and which may be concerned with various physiological functions of the plant.

Parenchyma cells retain the ability to divide even when mature. They also play an important role in wound recovery and regeneration. Mature parenchyma cells may resume meristematic activity when their environment is artificially changed. It has been shown recently that a small group of parenchyma cells, or even one such cell, when grown in proper culture media, may produce whole plants which will flower and produce viable seeds.

Phyiogenetically the parenchyma of the primary body is considered to be a primitive tissue as the lowest multicellular plants consist of parenchyma only. Ontogenetically parenchyma may also be considered primitive as its cells are morphologically similar to those of meristems. Large portions of the plant, such as the pith, all or most of the cortex of the root and shoot, the pericycle, the mesophyll of the leaf, and the fleshy parts of fruits, consist of parenchyma.

Shape

The shape of the paranchymatous cells differs greatly. Many parenchyma cells are polyhedral and their diameter in the different planes is more or less equal but many other shapes are common.

Elongated parenchyma cells are found in the palisade tissue of the leaf, in the medullary rays, etc.; lobed cells are found in spongy mesophyll and in the palisade parenchyma of *Lilium*; and in the mesophyll of the Xanthorrhoeaceae the parenchyma cells have folds or projections.

Stellate parenchyma cells are found in the stems of plants with well-developed air spaces, such as *Scirpus* and *Juncus*, for example. In *Juncus*, according to Geesteranus, the stellate pith cells differentiate ontogenetically from a mass of cubo-octahedral cells which are arranged in vertical rows on their hexagonal faces. The mechanical stretching of the pith, which is mainly in a radial direction as a result of the growth of the surrounding tissues, as well as the special arrangement of the intercellular spaces, causes the development of the characteristic arms of the cells.

Parenchyma cells with inner wall protuberances occur in various anatomical structures concerned with short distance transfer of solutes, such as in nectaries, salt glands and vascular parenchyma. These cells were termed transfer cells by Gunning *et al.*

Medium-sized polyhedral parenchyma cells usually have 14 faces; the number of sides is less in the smaller cells and greater in the larger cells. The number and size of the intercellular spaces also affects the number of faces of the polyhedron as the presence of intercellular spaces reduces the planes of contact between the cells. The polyhedral shape of the parenchyma cells is the result of numerous factors among which are pressure and surface tension.

Mature parenchymatous tissue may be tightly packed and without intercellular spaces or it may have a well-developed system of intercellular spaces. For example, the parenchyma of the endosperm of most seeds is devoid, or almost so, of intercellular spaces, while in the stems and leaves of hydrophytes the intercellular spaces reach maximal development.

The tissue with the prominent intercellular spaces occurring in plants, growing in waterlogged soils and aquatic habitats, is important in aerating the plant, and is called *aerenchyma*. The development of intercellular spaces is either *schizogenous* or *lysigenous*. The schizogenous development of an intercellular space takes place as follows: at the time when the primary wall is formed between two new cells the middle lamella between the two new cell walls comes into contact only with the primary wall of the mother cell and not with the middle lamella between it and the neighbouring cells. A small space develops where the new middle lamella, comes into contact with the mother cell wall. That portion of the mother cell wall opposite this small space disintegrates and so forms the intercellular space which can be enlarged by the formation of a similar space in the neighbouring cell.

The intercellular space is lined with the substance of the middle lamella. These intercellular spaces may be further enlarged by divisions of the surrounding cells in a plane perpendicular to the circumference of the space. The resin ducts of the Coniferae, the secretory ducts of the Compositae, Umbelliferae, *Hedera helix, Eucalyptus* are formed schizogenously, Lysigenous intercellular spaces are formed by the disintegration of entire cells. Examples of lysigenous intercellular spaces are the large spaces in water plants, in the roots of some monocotyledons and in the primary resin ducts of *Mangifera indica*.

Structure

Most parenchyma cells usually have thin primary walls, but parenchyma cells with thick primary walls also exist. Certain parenchyma storage cells as, for example, the endosperm of *Phoenix, Diospyros,* Coffea, and *Asparagus,* have very thick walls in which hemicellulose, which serves as the reserve substance accumulates. The walls of these cells gradually become thinner during germination. Parenchyma cells with relatively thick and lignified secondary walls are common, especially in the secondary xylem.

The internal structure of the parenchyma cell varies according to its function. Parenchyma cells which take part in photosynthesis contain chloroplasts and then the tissue they form is termed *chlorenchyma.* In the photosynthetic parenchyma there are usually single or numerous vacuoles. Certain parenchyma cells contain leucoplasts. Parenchyma cells may serve to store different reserve materials which may be found in solution in the vaculoes, or in the form of solid particles or liquid in the cytoplasm. Sugars or other soluble carbohydrates and nitrogenous substances may be found in the cell sap. Amides, proteins, and sugars are found dissolved in the cell sap, as, for example, in the roots of the sugar-beet and in the bulb scales of Allium *cepa*. Starch, protein, oils, and fats occur in the cytoplasm in the form of small particles.

Protein and starch grains are found in the cytoplasm of the cells of the cotyledons of many species of the Leguminosae, and protein and oils are found in the endosperm of *Ricinus communis* and the cotyledons of *Glycine max*. In the parenchyma cells of the potato tuber, for example, amides and proteins are found in the cell sap and starch in the cytoplasm. Starch is the most common reserve material in plants and it is found in the endosperm, cotyledons, tubers, fruits, xylem and phloem parenchyma, the cortex, etc.

In succulent plants, parenchyma cells that store water are present. Such cells are usually large, thin-walled, and have only a thin layer of cytoplasm, and they are devoid of or contain very few chloroplasts. The water-storing cells have a large vacuole which contains somewhat mucilaginous sap. The mucilaginous substances apparently increase the water-holding capacity of the cell, and they are also present in the cytoplasm and wall.

The parenchyma cells of different storage organs contain water as well as reserve substances, as, for instance, in the potato tuber which supplies water to the developing parts of the plant at the start of the sprouting process. Many parenchyma cells contain tannins, and such cells may be scattered throughout the plant or they may form continuous systems.

Most of the tannins are found in the vacuoles. Tannin-containing cells retain the ability to divide and grow as do parenchyma cells devoid of tannins. Mineral substances can be found in various crystalline forms in parenchyma cells. Some such cells may remain viable after the formation of the crystals but others die. The elongated rectangular cells with secondary walls, which often occur below the epidermis of many conifer leaves or in the xylem, can be considered as parenchymatous elements. For these cells the term *sclerified* parenchyma cells is suggested.

Sclerenchyma

Sclerenchyma is a tissue composed of cells with thickened secondary cell walls, lignified or not, whose principal function is support and sometimes protection. Sclerenchyma cells exhibit elastic properties, unlike collenchyma cells which exhibit plastic properties. Sclerenchyma cells may differ in shape, structure, origin, and development. Many transitional forms exist between the various cell shapes, and thus it is difficult to classify the different types of sclerenchyma.

Sclerenchyma is divided into *fibres* and *sclereids.* Fibres are usually defined as long cells and sclereids as short cells. This definition is not sufficient, as very long sclereids exist and relatively short fibres can be found. It has been suggested that the pits are very narrow, have round apertures, and are more numerous in sclereids.

The pit cavities of the sclereids very often have the form of branching canals which is a result of the increase in thickening of the wall. An attempt was made to distinguish between fibres and sclereids on the basis of the origin of the elements. Sclereids develop from parenchyma cells whose walls become secondarily thickened, whereas fibres develop from meristematic cells and so they are determined from their origin.

Other research, however, has shown that these definitions are also insufficient due to their inconstancies. Not only is it difficult to distinguish between the different types of sclerenchyma cells because of the existing transitional forms, but it is also somewhat difficult to distinguish between sclerenchyma and parenchyma as there are parenchyma cells with thick secondary walls, such as the xylem parenchyma. Fibres occur in different parts of the plant body. They may occur singly as idioblasts (e.g. in the leaflets of *Cycas*), but more usually they form bands or a network or an uninterrupted hollow cylinder.

Fibres are most commonly found among the vascular tissues but in many plants they are also well developed in the ground tissues. According to their position in the plant body, fibres are classified into two basic types— *xylary* and *extraxylary* fibres. *Xylary fibres* constitute an integral part of the xylem and they develop from the same meristematic tissues as do the other xylem elements. These fibres are of varied shape in spite of their common origin.

There are two main types of xylary fibres, i.e. *libriform fibres* and *fibre-tracheids,* are distinguished on the basis of wall thickness and type and amount of pits.

Libriform fibres resemble phloem fibres and they are usually longer than the tracheids of the plant in which they occur. These fibres have extremely thick walls and simple pits. Fibre-tracheids are forms intermediate between tracheids and libriform fibres. Their walls are of medium thickness-not as thick as those of the libriform fibres but thicker than those of the tracheids. The pits in fibre-tracheids are bordered but their pit chambers are smaller than those of tracheids.

In fibre- tracheids and sometimes also in libriform fibres the pit canal is elongated and the inner pit aperture usually becomes slit-like as a result of the thickening of the wall. In fibre-tracheids, therefore, the length of the pit aperture usually exceeds the diameter of the pit chamber. In both libriform fibres and fibre-tracheids, the inner pit apertures of a pit-pair are usually at right angles to each other.

Another type of fibre present in the secondary xylem of dicotyledons is the *gelatinous* or *mucilaginous fibre*. In such fibres the innermost layer of the secondary wall contains much α-cellulose and is poor in lignin. This layer, termed the *G-layer,* absorbs much water and may swell so as to fill the entire lumen of the fibre. On drying, these layers shrink irreversibly. The G-layers were found to be relatively porous and less compact than the adjacent outer layers.

Extraxylary fibres occur elsewhere in the plant other than among the xylem elements. They occur, for instance, in the cortex or they may be closely related to the phloem elements. In the stems of many monocotyledons, the extraxylary fibres occur in an uninterrupted hollow cylinder in the ground tissue, and they may be situated at various distances inside the epidermis and may even surround the outermost vascular bundles.

Commonly in the monocotyledons, the fibres form sheaths around the vascular bundles. Such fibres develop partly from the procambium and partly from the ground tissue. In the stems of climbing and certain other dicotyledonous plants, such as *Aristolochia* and *Cucurbita,* fibres are found on the inside of the innermost cortical layer and on the periphery of the central cylinder. These fibres are not developmentally connected with the phloem and so they were termed, by many workers, *pericyclic fibres.*

As a result of ontogenetic studies in stems of some plants. It has been concluded that the so-called pericyclic fibres of the investigated species develop from the procambium and thus represent primary phloem fibres.The above classification into xylary and extraxylary fibres is not always applicable as there are fibres, such as the *septate fibres,* which are found in the *xylem* and the phloem even of the same species, e.g. in *Vitis* where they are very common. These fibres are characterized by the presence of internal septa and, usually, of a living protoplast.

As seen in septate wood fibres of *Ribes sanguineum* the internal septa result from mitosis in lignified cells. The septum does not fuse with the fibre wall but becomes broadened, at the place of contact with it. The fibre tips run out as pointed ends when viewed in longitudinal section of the fibre. The septum consists of a middle lamella and two primary wall-like layers interrupted by numerous plasmodesmata. Septate fibres may contain starch and oils and, therefore, are thought to have a storage function. They may also contain resins and sometimes crystals of calcium oxalate. Septate fibres are present in the secondary xylem of many dicotyledons.

Non-vascular septate fibres occur in some monocotyledons, e.g. in Palmae and in Bambusoideae.

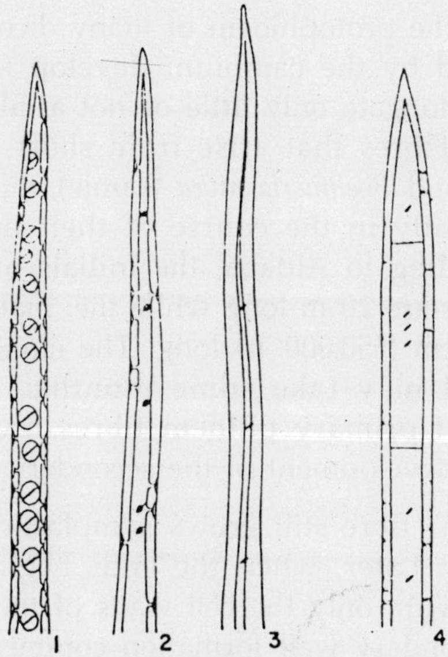

Fig. 1. Tips of elements from the secondary xylem of Quercus ithaburensis. (1) Tracheid (2) Fibre-tracheid. (3) Libriform fibre. (4) Septate fibre of vitis.

Mature fibres have well-developed, usually lignified secondary walls, which are sometimes so thick as to obscure the lumen of the fibre. Lamellae can be distinguished in these walls. In *Linum,* for example, each lamella is 0.1-0.2 μ thick as seen in a cross-section of the fibre. Mention should be made here of those elongated cells that sometimes occur in the secondary xylem and whose secondary walls are equal in thickness to those of the xylem parenchyma.

Fibres

Fibres develop from different meristems, such as the procambium, cambium, ground meristem, and even from the protoderm, as in certain species of the Gramineae and Cyperaceae. Fibres may also develop from parenchyma cells, e.g. in the protophloem of many dicotyledons. The fibres formed by the cambium develop from fusiform initials and elongate only little or not at all during their maturation. Fibres that arise from short initials, as in *Linum* (flax) and *Boehmeria nivea* (ramie), must necessarily elongate greatly in the course of their maturation. In ramie, according to Aldaba, the initials of the primary phloem fibres are 20 m long while the, mature fibres can be up to 55 cm (550,000 μ) long. The elongation is very gradual and may take some months. This gradual elongation of primary phloem fibres involves a very complicated development of the secondary wall.

While the fibre still grows symplastically, the wall remains thin. Later, when the ends begin to grow by intrusive growth, only the cell walls of the ends remain thin and secondary wall formation commences from the middle of the fibre in those parts of wall which have ceased to elongate. In *Linum* and ramie it has been found that this process is gradual so that new lamellae of the secondary wall are added centripetally in the form of cylinders which are open at both ends. At the same time the first-formed lamellae continue to elongate towards the fibre ends which they reach only when the fibre ceases to elongate.

The upper ends of ramie fibres continue to grow for a longer period than the basal ends. Sometimes not all the lamellae reach the actual fibre end and in some fibres chambers may be formed in the terminal portions by the ingrowth, towards the cell lumen, of these lamellae. The lamellae of the primary phloem fibres, or at least of the immature fibres, are often not strongly attached to one another. This feature is easily demonstrated during the cross-sectioning of such material when the different layers become torn one from the other.

In short fibres, such as those found in *Agave, Sansevieria,* and *Musa textilis,* whose total length is not more than a few millimetres, all portions of the cell wall grow at the same rate. Differences exist in the manner of growth of the fibres in the primary body and of those in the secondary body.

The initials of the primary fibres appear early, before the organ in which they occur has elongated, and so they may grow in length symplastically together with the neighbouring cells which continue to divide. The symplastic growth is augmented by intrusive and gilding growth of the ends which thus penetrate between the surrounding cells. The initials of the secondary fibres develop in organs that have ceased to elongate and therefore the growth of secondary fibres can be intrusive only. This is apparently the reason why the primary fibres are usually longer than the secondary fibres of the same plant. Thus it was found in ramie that the average length of the primary phloem fibres is 164.6 mm while that of the secondary phloem fibres is 15.5 mm.

Electron microscopic studies of wood fibres have shown that during secondary wall formation the cytoplasm was in general conspicuously vacuolated, although the vacuolation did not extend to the cell tips. The nucleus was large and the organelles which were generally few in number appeared more crowded near the tips. The ER frequently appeared to lie parallel to the wall surface.

The xylary fibres differ in shape, size, thickness of wall, type, and amount of pits. It is assumed, from the evolutionary point of view, that fibres have developed from tracheids. This assumption is supported by the fact that many transitional forms between these two types of elements are found in some angiosperms, as, for example, *Quercus,* spp.

From the many transitional forms that have been distinguished it appears that the following changes have taken place during the course of the evolution of fibres from tracheids. The wall has become thickened, the number of pits and the size of the pit chamber has been reduced, leading to the eventual disappearance of the bordered pit, and the cells have become shortened. This assumed shortening of the fibres refers to the shortening of the initials of the fibres in the cambium and not to, the mature fibres.

The term *fibre,* as used in industry, does not generally have the same meaning as that defined by botanists. For instance, the commercial fibres of *Linum, Boehmeria,* and *Corchorus* are, in reality, a bundle of fibres, and those from monocotyledonous leaves, such as from *Agave, Musa tex*tilis, and others, are usually the vascular bundles with the surrounding sheaths of fibres.

Commercial fibres are divided into two types—hard fibres and soft fibres. Hard fibres are those which have a high lignin content in the walls, and are of a stiff texture. Hard fibres are obtained from monocotyledons. Soft fibres may or may not contain lignin, they are flexible and elastic, and are of dicotyledonous origin.

The best-known plants from which hard fibres are produced are different species of *Agave,* especially A. *sisalana* (sisal), *Tillandsia usneoides* (Spanish moss), Musa *textilis* (abaca), *Furcraea gigantea* (Mauritius hemp), and *Phormium tenax.* Soft fibres are mainly produced from *Linum usitatissimum* (flax), *Cannabis sativa* (hemp), *Boehmeria nivea* (ramie), *Corchorus capsularis* (jute), Hibiscus *cannabinus* (kenaf), and *Ceiba pentandra* (kapok). The fibres of cotton, which are produced from the indumentum of seeds, represent the most important commercial fibres in use today.

Fibres are also classified according to their use:

i) textile fibres which are used in the manufacture of fabrics;

ii) cordage fibres:

iii) brush fibres such as are used in the manufacture of brushes and brooms; and

iv) filling fibres such as those used for stuffing upholstery, mattresses and life-belts, caulking (barrels, plumbing), and reinforcing (wall plates, plastics).

In the textile industry the principal fibre used is cotton and, in smaller amounts, flax, ramie, and hemp. For coarser fabrics, such as sacking and bagging, jute is principally used, and cotton, flax, hemp, and a few other hard fibres are used to a lesser extent.

For the manufacture of twine, jute, cotton, hemp and, to a lesser extent, flax and several hard fibres are used. Ropes and binder twines are manufactured from hard fibres, such as those of *Musa textilis* (abaca) and *Agave spp.* (sisal), and to a small extent from cotton and other soft fibres. Brushes and brooms are made from *Agave* fibres, fibres from the stems and leaves of the Palmae and the inflorescences of *Sorghum vulgare,* among others.

As filling fibres, the fibres of *Ceiba pentandra* (kapok) cotton, jute, the fibres of *Tillandsia usneoides,* several hard fibres, and others are used. For caulking fibres, hemp, jute and sisal are used. Fibres are also used in the paper industry; depending on their physical and chemical properties different types of paper may be made.

The shape of the fibre cell, its length, and wall structure are of importance in the fibre industry. Special attention is paid to the length of the fibre, the extent to which neighbouring fibres overlap, and how they are joined to one another. The orientation of the cellulose units in the wall has an important effect on the physical properties of wood and commercial fibres.

Elasticity and heat conductivity increase as the degree of orientation parallel to the length of the fibre increases. Swelling is much greater in a direction perpendicular to the cellulose molecules than it is in a direction parallel to them; this is the case even in water, where swelling is almost exclusively intermicrofibrillar. Commercially, fibres are ranked according to durability,, tensile strength, length of the strands, fineness, uniformity, and elasticity.

Sclereids

Sclereids occur in many different places in the plant body. In many plants they occur as hard masses of cells within soft parenchyma tissue. Certain organs, such as the shell of walnuts and many other stone and seed coats, are built entirely of sclereids.

In many plants sclereids appear as idioblasts, i.e., as cells which are readily distinguished from the surrounding cells of the tissue by their size, shape, and the thickness of their wall. Idioblastic sclereids are very variable in shape.

Sclereids of peculiar shape are found in the leaves of various plants, e.g. in *Gnetum, Camellia, Trochodendron, Nymphaea, Cyathocalyx, Desmos, Phaeanthus, Horsfieldia, Salvadora, Monstera,* and *Olea* and in aerial roots of *Monstera.* In the leaf mesophyll the sclereids may either by irregularly diffused *(diffuse sclereids),* e.g. in *Trochodendron, Olea, Osmanthus,* and *Nymphaea;* or the sclereids may be confined to the vein endings *(terminal sclereids),* e.g. in *Mouriri, Boronia, Arthrocnemum, Capparis spinosa* and *Salvadoraceae* species. The elongated sclerenchyma cells occurring in the mesophyll of the Myrtaceae and some other families are often referred to as fibres. In some plants, e.g. the Magnoliaceae, sclereids are found in almost all organs: in the cortex and pith of the stem, leaves, stipules, floral appendages, receptacles, fruits, and rarely in roots.

Among the types of idioblastic sclereids there is also a type of cells that simulate tracheids, *tracheoid idioblasts*. Such tracheoids are found, for instance, in *Salicornia* species and in inflorescence stalks of some orchids. Tschirch suggested the division of sclereids into four types:

i) *Brachysclereids* or *stone cells,* which are more or less isodiametric in form; such sclereids are usually found in the phloem, the cortex and the bark of stems and in the flesh of such fruits as pears (*Pyrus communis)* and quinces *(Cydonia oblonga)*;

ii) *Macrosclereids,* which are rod-shaped sclereids, such sclereids often form a continuous layer in the testa of seeds, e.g. in the seeds of the Leguminosae;

iii) *Osteosclereids,* which are bone or spool-shaped sclereids, the ends of which are enlarged, lobed, and sometimes even somewhat branched; such sclereids are mainly found in seed coats and sometimes also in the leaves of certain dicotyledons;

iv) *Asterosclereids,* which are variously branched and often star shaped; such sclereids are mainly found in leaves.

A fifth type of sclereids, termed *trichosclereids,* was suggested by Bloch. These are very elongated, somewhat hair-like, and regularly once-branched sclereids.

Additional terms for sclereid types, such as *filiform, fibriform, columnar,* and *polymorphic,* which occur in leaves, are used by various authors.

Typical brachysclereids develop from parenchyma cells by secondary thickening of the cell wall. This secondary wall is very thick, and numerous concentric layers and branched pits can usually be distinguished in it. During the process of wall thickening the inner surface of the wall decreases and pits which start to develop on the outside of the secondary wall are brought together.

The physiological reasons for the sclerification of the parenchyma cells are not known, but Bloch thought the fact that stone cells often appear close to wound tissues suggests that they develop in response to some physiological disturbances. In the bark the change of many parenchyma cells to sclereids suggests that, in this case, the cause is the ageing of the tissue. Recent observations revealed that although in the wound meristem of injured leaves certain cells develop into sclereids, wounds have an inhibitory influence on the differentiation of sclereids in lamina of developing leaves which normally produce sclereids.

Another interesting example of the development of stone cells is in the continuous cylinders of the phloem fibres. When stems of plants with such cylinders grow in width, neighbouring parenchyma cells penetrate into the spaces formed between the fibres of the cylinder. The parenchyma cells then divide and become sclerified and so close the gaps formed in the cylinder.

Sclereids are usually described as non-living cells when mature, but it has been seen that the protoplasts may remain viable throughout the life of the organ in which the sclereids are found. In non-deciduous leaves and in certain stems the life of the sclereids may sometimes be 4-5 years. The protoplast in the stone cells in the fruits of the pear and quince also remains alive for relatively long periods.

According to Alexandrov and Djaparidze, during the ripening of the quince fruit, the stone cells undergo a process of delignification, which they believe is an indication of the enzymatic activity of the protoplast of the stone cell itself. The structure of the fibre wall has been investigated comparatively thoroughly and emphasis laid on the wall structure of fibres of economic value. Attention has also been paid to the ontogenetic and phylogenetic development of fibres. Cell growth, especially intrusive growth, can be well studied in the course of fibre development.

Collenchyma cells may often become sclerified during the maturation of the organs in which they occur. This fact emphasizes the view of the close relationship between these two tissues. Because of the great variability in the form of fibres and because of the existence of many transitional forms, fibres serve as favourable material for the study of the evolution of an element or part of it, as, for instance, the evolution of the pit.

Substitute and septate fibres have, during the course of evolution, become strikingly different from the typical fibre form and should, actually, be classified as parenchyma cells with secondarily thickened walls. A substitute fibre resembles an elongated parenchyma cell and a septate fibre a longitudinal series of parenchyma cells derived from a single mother cell in which secondary wall lamellae develop before the cell divisions are completed.

The libriform fibres and the fibre-tracheids have, till recently, generally been described as non-living cells devoid of protoplasts and were regarded as having only mechanical function or, at the most, as playing a small role in water conduction in addition to the tracheary elements. However, it now appears, in the light of recent research, that the libriform fibres and even fibre-tracheids of the sap wood of many woody plants contain living protoplasts. Therefore we should begin to consider fibres not only as supporting elements but also as elements that doubtless fulfil various other important physiological functions.

Xylem

Xylem is a complex tissue as it consists of several types of cells. The vascular system, and especially the xylem, has been used for the classification of a large group of plants. The term *vascular plants* was first used in 1917 by Jeffrey. Later, the term *Tracheophyta* has been introduced to cover this group of plants which comprises the Pteridophyta and Spermatophyta. The term *Tracheophyta* has been derived from the xylem, and not the phloem, because of the firm and enduring structure of the tracheary elements. These elements have thick, hard walls and so can be distinguished more easily than the phloem elements.

The xylem is more readily preserved in fossils and so can be identified more easily. The most important cells are the *tracheary elements* which are the non-living cells that are principally concerned with the transport of water and which also, to a certain degree, have a supporting function.

Fibres are present in the xylem where they are mainly concerned with the strengthening of the plant body. Sometimes sclereids may also be present. Parenchyma cells which have storage and other functions also occur in the xylem. The xylem of some plants contains resin ducts. The xylem and phloem elongate in developing organs by the continual differentiation of new elements produced by the procambium, which itself is continuously produced by the apical promeristem.

The xylem produced by the procambium in the primary body is called the *primary xylem.* In many plants, after the completion of the formation of the primary body, secondary tissues are developed. The xylem that is produced as a result of the activity of the vascular cambium is called the *secondary xylem.*

Tracheary elements

The tracheary elements are of two types—*tracheids* and *vessel members.* Since then much work has been devoted to the investigation of the structure, shape, function, ontogeny, and phylogeny of these elements. The main difference between tracheids and vessel members is that the former are not perforated while the end walls of the latter are perforated.

A vessel, which is also termed a *trachea,* is built of numerous vessel members that are joined one to the other by their end walls. Vessels are terminated by a vessel member of which the proximal end wall is perforated, whereas the distal end wall is not, i.e. the distal parts of a vessel are tracheid-like.

In a radial longitudinal section of vascular bundles it can be seen that the tracheary elements differ one from the other in the shape and structure of the secondary wall. In many plants the secondary wall thickening of the first-formed xylem (protoxylem) is *annular* or *helical.* The helical thickening may be single, or more than one helix may be present in a single element. The rings or helices may be arranged in a loose or a dense manner.

From an ontogenetic viewpoint, the annular elements precede the helical elements. In later-formed tracheary elements (in early metaxylem) the helical bands become joined in certain areas, giving rise to a ladder-like type thickening; such thickening is termed *scalariform thickening*. In tracheary elements formed at a still later ontogenetic stage the wall thickening is in the form of a network, i.e. *reticulate thickening*. When the openings in the secondary wall of such a network are elongated in a direction perpendicular to the longitudinal axis of the clement, the thickening is termed *scalariform-reticulate*. In the ontogenetically most advanced elements, the secondary cell wall is interrupted only at the pits; such elements are termed *pitted elements*.

Pitted elements are characteristic of the late primary and of the secondary xylem. Not all the above types are always found in a single plant. On the other hand, intermediate types not mentioned above can be found, as well as combinations of more than one form of thickening which may occur in a single element.

The annular and helical wall thickenings may vary in thickness and certain helices are so deeply grooved on their inner surfaces as to appear double. In some cases the helical thickening is joined by a narrow strip to the primary wall. In general, it has been shown that the range of variation in the pat tern of wall thickening of the tracheary elements is very extensive. The pits in pitted tracheary elements are bordered. The well-developed bordered pit-pairs which are usually present between two tracheary elements are termed *intervascular pits*.

Between tracheary elements and fibres there may be only a few small pits or even none at all. Between tracheary elements and parenchyma cells the pit-pairs are mostly half-bordered, i.e. bordered on the side of the tracheary element and simple on the side of the parenchyma cell. When the bordered pits are transversely elongated and are arranged in longitudinal rows along the element, the pitting is termed *scalariform pitting*.

Circular and elliptical pits are arranged in horizontal or diagonal rows. The former arrangement is called *opposite pitting,* and the latter *alternate pitting.* On the inside surface of a pitted secondary wall a helical thickening may develop. In the Ophioglossales, the Ginkgoales, the Coniferales, and the Gnetales no scalariform pitted elements are found. In plants of these orders bordered pits, which are similar to those found in the secondary tracheary elements of the same plant, are found on the reticulate and helical thickenings of the primary xylem.

Cruger observed that in the positions where the thickenings of the secondary wall will develop, strips of actively streaming cytoplasm appear. Similar conclusions were reached by Barkley who believed that the position of the cytoplasmic strips is determined by the position of rows of vacuoles. A similar phenomenon was also observed by Sinnott and Bloch who studied the development of tracheary elements from parenchyma cells during the regeneration in the vascular bundles of *Coleus.* Several workers have suggested that Golgi bodies are involved in the formation of the wall thickenings. There are indications that the endoplasmic reticulum may also be con cerned with this process.

Recently it has been observed that preceding the formation of the wall thickenings of the tracheary elements microtubules become grouped in bands between small, regular corrugations in the wall. In later stages many vesicles, probably derived from Golgi bodies, collect at the regions occupied by the microtubules.Labelling indicated that both the endoplasmic reticulum and Golgi bodies were associated with the wall incorporation of lignin precursors. Cronshaw has noted that microtubules which are often seen in the cytoplasm close to the cell surface and, according to Ledbetter and Porter, may govern cytoplasmic streaming or exert an influence over the disposition of cell-wall material, are oriented in the same direction as the cellulose microfibrils of the developing wall thickenings of tracheary elements of the protoxylem of the *Avena* coleoptile. Barnett, who studied tracheid differentiation in *Pinus radiata,* did not find significant numbers of microtubules during the main phase of secondary wall formation. Their number, however, increased much in the final stage of secondary wall formation, when lignification of the secondary wall begun.

After completion of the elaborate secondary wall thickenings and their lignification, the protoplast and certain parts of the cell wall enter the stage of lysis. It has been suggested that autolysis of the cytoplasm occurs when the latter is exposed to the vacuolar sap containing hydrolytic enzymes after a rupture of the tonoplast. Hydrolysis affects also the noncellulosic components of those parts of the primary wall that are not covered by lignified secondary wall.

The residual material of the hydrolysed primary wall forms a coarse net of fibrils. According to Czaninski, in *Robinia* at the end of the differentiation of the vessel element a tertiary wall lamella may be deposited. Scott *et al.* described it suberization of the vessel wall in *Ricinus* after the death of the protoplast. According to Bierhorst and Zamora the tracheary elements which mature later than the helical ones go through a two-phase wall deposition process.

In the first phase the helical framework is formed and during the second phase secondary wall material is deposited between the gyres of the helix. The functional significance of the different types of wall thickenings in the tracheary elements is not clear. It is possible that the exclusive appearance of annular and spiral thickening in elements in those organs that are still elongating has some connection with the rapid increase in length of the organ. Investigations using X-rays together with the regulation of light which altered the rate of stem elongation proved this assumption. Goodwin and Smith and Kersten saw that if stem elongation is inhibited the production of annular and spiral is reduced or stopped and pitted vessels.

The development of vessel members with wall thickenings that are intermediate between protoxylem and metaxylem, from a file of procambial cells which is continuous with typical protoxylem elements, occurs at the end of intercalary growth of the internode of *Cyperus alternifolius*. This supports the view that the extent of elongation of an organ determines the wall pattern of the tracheary elements maturing in the organ.

Vessels develop from meristematic cells—procambial cells in the primary xylem and cambial cells in the secondary xylem. The vessel members may or may not elongate prior to the thickening of the wall but they usually widen in this stage of development. During the stage of the growth of the, meristematic vessel element, the nucleus becomes polyploid and increases in volume.

Much attention has been paid by workers studying the ontogeny of vessels to the end walls in which the perforations develop. The end wall occupying the future site of the perforation does not become covered with a secondary wall layer, but it does become thicker than the primary wall elsewhere.

On the basis of a cytochemical study of the differentiation of the end wall of secondary vessels of *Populus italica* and *Dianthus caryophyllus* it was suggested that this wall consists primarily of non-cellulosic polysaccharides and of only very little cellulose. The manner of removal of the end wall may differ in different species and in vessel elements of the primary or secondary xylem. So, for instance, it was reported that, in the central vessel of the root of *Hordeum vulgare* lysis begins along the margin of the end wall. In the secondary vessel elements of *Knightia excelsa* small holes develop in the end wall before final disappearance. A similar process occurs in *Quercus rubra*. In *Pisum sativum* and in *Dianthus caryophyllus* lysis of the end wall begins in the centre.

The xylem holds an important position in the study of plant tissues as the structure of its elements is of extreme importance in taxonomy and phylogeny. More attention has been paid to the phylogenetic development of xylem than to any other tissue. The tracheid is the only tracheary element found among Pteridospermae, in fossil Spermatophyta, in most of the lowest vascular plants of today, and in nearly all the Gymnospermae.

It is commonly accepted that vessel members have developed from tracheids. Vessel members occur in the following diverse groups of plants: in the most advanced gymnosperms, the Gnetales; in dicotyledons except for the lower-most taxonomic groups; in monocotyledons; in the fern, *Pteridium;* in certain species of *Selaginella;* in *Equisetum*, and in the root of some species of *Marsilea*. Vessels developed independently, by parallel evolution, in each of these groups.

Frost clearly defined some of these fundamental logical assumptions while trying to establish the origin of the vessels in the dicotyledons. The following are the principal assumptions used by Frost.

i) *The association method:* This method states that if it is possible to determine which of two structures is the more primitive, and if it is assumed that the two structures have a direct genetic relationship, it will be possible to conclude that the primitive condition of the more advanced structure will be similar to the general condition of the primitive structure. If there is not much similarity then the assumption of direct genetic relationship is not correct or the elements in question are, apparently, so far separated in the scale of evolution that the primitive form of the advanced element has been lost. Therefore, with reference to tracheary elements, if it is assumed that the tracheids are more primitive than the vessel members and that the two elements have a direct genetic relationship, then it must be concluded that the most primitive vessel members will be those that are most similar to tracheids.

ii) *The correlation method:* By this method it is assumed that in a certain homogeneous tissue, as, for instance, the secondary xylem, there will exist a statistically significant correlation between the degrees of specialization of the main characteristics of a structure in a large random sample, i.e. the various features have undergone evolutionary changes simultaneously. Therefore, features occurring together with those features that are defined as being primitive by the association method are themselves primitive, and those that occur together with features defined as being advanced are advanced. It is necessary to bear in mind that such correlations express only the general trends of development and that exceptions exist. The development of some features may be delayed and of others advanced. The investigations of these exceptions can indicate the lines of secondary specialization which only become clear after the principal lines of development have been determined. In relation to the vessel members, if great length is a primitive feature then all other features that are found in correlation with great length are also primitive features.

iii) *The sequence method:* This method deals with the reconstruction of the evolutionary variability on the basis of the variation, as seen in living forms. These variations can be seen ontogenetically or by the comparison of different plants belonging to a single taxonomic group. The contribution of this method to the problem of the origin of vessels has been the determination of the origin of vessel members from tracheids with scalariform pitting. Typical tracheids of this type appear only in the secondary wood of those dicotyledonous genera that have no vessels, e.g. genera of the Winteraceae, Monimiaceae, Chloranthaceae, and Tetracentraceae, but are completely absent from the secondary xylem of angiosperms that contain vessels. In some trees and large shrubs of various primitive dicotyledonous families, the vessel members of the secondary xylem are similar in size, angular cross-section, pitting, and thin secondary walls to tracheids with scalariform pitting. It is important to mention that the scalariform pitted tracheids have served as the origin not only of vessel members but also of tracheids with circular bordered pits and apparently also, indirectly, of fibre-tracheids and libriform fibres.

The complete or almost complete absence of primitive tracheids with scalariform pitting in the Angiospermae is related to their development, in the process of phylogeny, into vessel members or into tracheids with a more advanced form of pitting.

structural features of tracheary elements

The following structural features of the angiosperm tracheary elements are those that are used as a basis for the study of their evolution.

— *Length:* The tracheids are long cells whose average length reaches 4.35 mm in *Trochodendron,* a vessel-less dicotyledon. Their average length as calculated from many hundreds of measurements made in monocotyledons is 5.07 mm. In monocotyledons, according to Cheadle, vessels can be divided into four groups according to their degree of specialization. The average lengths of the vessel members in these four groups are 3.96 mm, 2.58 mm, 1.47 mm, and 0.76 mm. As the vessel members are shorter than the tracheids the shorter the vessel member, the more advanced it is considered to be.

— *Diameter:* The diameter of the tracheid is smaller than that of the vessel member.

— *Thickness:* The wall of a tracheid is thin and is of equal thickness over the entire circumference. This feature is also seen in primitive vessel members.

— *Perforation plates:* Those scalariform perforation plates that are long, oblique and with numerous perforations are considered the most primitive and the simple, horizontal perforation plates the most advanced.

— *Shape in cross-section:* The shape of the tracheids and the primitive vessels in cross-section is angular, while that of advanced vessel members is circular or nearly so.

— *Type of pitting:* In the dicotyledons scalariform pitting in vessel members is considered to be primitive. The structure and arrangement of pits developed from scalariform pitting, through intermediate forms in which scalariform pits occur together with circular or elliptical pits, to forms with only circular or elliptical pits. Of this advanced type of pitting, that in which the pits are arranged in parallel rows, i.e. opposite, pitting, is more primitive than alternate pitting, in which the circular or elliptical pits are arranged along more or less helical lines. The appearance of the spiral thickenings on the inside of the secondary wall of the tracheary elements is evidence of advanced development.

Vessels in the monocotyledons first appeared in the roots and later in the stems and leaves. The specialization of the vessels followed the same pattern. Phylogenetically, the vessels first appeared and became specialized in the late-formed metaxylem and progressed gradually, into the early formed metaxylem and finally into the protoxylem. Monocotyledons exist today that have, in the last-formed metaxylem of their roots, only the most primitive vessels the perforation plates of which are scalariform and which contain more than 100 parallel perforations.

A few monocotyledonous families with only aquatic species are known to include plants that lack vessels completely in all their organs. This feature, however, may be a secondary one. The tracheary elements have developed during the evolution of the land plants. Two main functional trends have become evident during the course of the morphological evolution of these elements, i.e. the development of those structures that enhance rapid conduction, on the one hand, and of those that strengthen the elements, on the other hand. These two trends are antagonistic to a great extent because certain structures that increase the efficiency of conduction tend to weaken the cells and vice versa.

However, during the course of evolution, structures have been developed that have, to various extents, resolved these two trends. Elements with such wall thickenings give support to the mature stem. The absence of living protoplasts in the tracheary elements, the development of elongated tracheids and the occurrence of vessel members are all features that increase the efficiency of water conduction. The bordered pit-pairs which are characteristic of the tracheary elements, are, as has been shown by Bailey, well adapted to their function and they combine the two above-mentioned trends.

The area of the pit membrane is comparatively large and so the passage of water is fairly easy and, on the other hand, the extent of the development of the secondary wall is maximal because the secondary wall overarches the pit membrane in such a manner that the pit membrane remains comparatively large whereas the pit aperture is very small. This feature greatly strengthens the tracheary elements.

In tracheids more rapid conduction is obtained by the elongation of the cells, the increase in diameter of the lumen and in the number of pits and the reduction of wall thickness. Strengthening of the tissue is brought about by the shortening of the cells, narrowing of the lumen, increase in wall thickness and the reduction in the number of pits.

In the secondary xylem of conifers, for instance, the early wood is more adapted for efficient water conduction, and the tracheids of the late wood for support. Conduction is further facilitated by the complete disappearance of the pit membranes in certain areas, resulting in the formation of vessel members. In the secondary xylem of certain primitive dicotyledons primitive vessel members, resembling scalariform pitted tracheids, and thick-walled, narrow tracheids with a few round bordered pits have been observed to occur side by side.

This phenomenon proves that the evolution of tracheids, in relation to function, was dichotomous, i.e. the scalariform tracheids evolved into vessel members which were better adapted to conduction, whereas the tracheids with round bordered pits are modified to give better support and, through various intermediate forms, give rise to the libriform fibres.

The structural evolution of the tracheary elements presents one of the most convincing examples of evolutionary development. However, although it is obvious from the great amount of data that this structural evolution has been accompanied by functional specialization, as yet almost nothing is known about whether or not a correlation exists between the types of tracheary elements and ecological conditions.

The sequence of the different types of tracheary elements and the numerous transitional forms are important features in the study of the origin of the Angiospermae and the phylogeny of the various taxonomic groups among them. It is still necessary, however, because of parallel and convergent evolution, to accumulate more data concerning the tracheary elements of the various species and to use such data together with morphological and structural data concerning other elements and tissues before any definite conclusions can be drawn.

Phloem

The phloem, together with the xylem, constitute the conducting system of vascular plants. The xylem functions principally in the conduction of water, and the phloem of products of photosynthesis. Similarly to the xylem the phloem also is a compound tissue. The important cells of the phloem are the *sieve elements* which serve for the conduction of the photosynthetic products.

Additional to these elements phloem contains typical parenchyma cells in which reserve substances are stored as well as specialized parenchyma cells, i.e. the *companion cells* and *albuminous cells,* which are connected with functioning of the sieve elements. Fibres, sclereids, and sometimes laticifers and resin ducts may also be found in phloem tissue. The primary phloem, similarly to the primary xylem, develops from the procambium. The primary phloem is divided into the *protophloem,* which develops from the procambium during an early ontogenetic stage and matures before the organ in which it is located has ceased its longitudinal extension growth, and the *metaphloem* which also develops from the procambium, but at a later stage of development.

The phloem in the stem is usually external to the xylem, but in many ferns and in different species of numerous dicotyledonous families, e.g. Asclepiadaceae, Cucurbitaceae, Myrtaceae, Apocynaceae, Convolvulaceae, Compositae, and Solanaceae, phloem is also present on the inside of the xylem. Phloem on the inside of the xylem is called *internal* or *intraxylary phloem* and it develops a little later than the external phloem. In certain families, such as the Chenopodiaceae, Amaranthaceae, Nyctaginaceae, Salvadoraceae, and others, phloem is also present within the secondary xylem. This type of phloem is called *interxylary phloem* or *included phloem.* The extensive literature on phloem was summarized by Esau.

Sieve elements

The most characteristic features of sieve elements are the *sieve areas* in the walls and the disappearance of the nucleus from the protoplast. The sieve areas are interpreted as being modified primary pit fields and they appear as depressions in the wall in which groups of pores are located. *Connecting strands,* the contents of pores, connect the protoplasts of the neighbouring sieve elements. Sieve areas can be distinguished from primary pit fields by the following two features:

i) In the sieve areas the connecting strands are much thicker than the plasmodesmata that occur in the primary pit fields;

ii) In the sieve area each pore usually contains a small cylinder of *callose* which surrounds the connecting strand.

The diameter of the pores of the more specialized sieve areas, which are present in the sieve plates, varies in the different species from a fraction of a micron to 14 μ. In *Spiraea vanhouttei* the diameter of the pores is less than 1 μ, while in *Pyrus malus* and *Pyrus communis* the maximum diameter is only a little larger than 1 μ, in *Cucurbita* spp. it is 10.3 μ, and in *Ailanthus altissima,* 14.3 μ. The presence of callose can easily be demonstrated by staining with aniline blue or resorcin blue. In ultraviolet light even minute amounts of callose stained with aniline blue give a lemon-yellow fluorescence. Callose is an unbranched β-1,3-linked glucan, it is not birefringent and shows no structure under the electron microscope.

Callose also occurs on the walls of fungal cells, in the germination tube of pollen grains, in cystoliths, and has even been shown to be present in the primary pit fields of epidermal cells. There is evidence that in response to injury, callose is formed and deposited very rapidly. In view of this phenomenon some workers have questioned the presence of callose in active sieve elements in intact plants. In any case, callose is generally present in active sieve elements of material fixed according to the usual methods and serves as a diagnostic feature of sieve elements.

In sieve areas that are not highly specialized the connecting strands are extremely thin and are nearly indistinguishable from plasmodesmata. In highly specialized sieve areas the connecting strands are very thick and they stain intensely. When the sieve element is young, the sieve area is thinner than the other portions of the cell wall and it appears as a depression on the inner surface of the wall.

As the element matures additional callose is laid down not only on the cylinders, which line the pores, but also on the surface of the sieve area between the pores, so that, finally, in old elements, the sieve areas do not appear as depressions but as raised portions above the surface. When the element ceases to function the connecting strands become very thin and may even disappear.

When sieve areas develop close to one another the callose masses from each sieve area may fuse to form a single mass. Large masses of callose which develop with the cessation of function of the elements are called *definitive callose.* Usually, with the complete disintegration of the protoplast of the element the callus peels away from the sieve areas. In most dicotyledons the sieve elements function during a single growing season, but in certain plants, such as *Suaeda, Tilia,* and *Vitis,* the sieve elements function for two or more years. In woody stems of *Suaeda* and other plants with included phloem the sieve elements remain active in the whole sapwood. In palm stems the sieve tubes in the metaphloem function throughout the life of the plant and no conspicuous structural features which might account for the long life of the sieve elements can be discerned.

In Tilia no distinct changes can be seen in the sieve elements with the start of the resting season, while in Vitis large amounts of callose accumulate in the autumn, disintegrate in the spring before the commencement of cambial activity, and so the sieve elements start to function for a second year. The density and arrangement of the sieve areas on the sieve elements exhibit the same degree of variation as does the pitting on the walls of the tracheary elements.

On the basis of the thickness of the connecting strands and the degree of development of the callose cylinders, sieve areas with different stages of specialization can be distinguished. In certain plants, such as the Coniferales, all the sieve areas of an element are equal, while in other plants, for instance, most of the angiosperms, some of the sieve areas are more specialized, i.e. they have more well-developed connecting strands and callose cylinders than those found in other areas.

These more specialized sieve areas are usually situated on the end walls of the elements which are horizontal or oblique to the longitudinal axis of the element. Those parts of the cell wall that bear such specialized sieve areas are called sieve plates. Apparently, the formation of the callose platelets takes place after the endoplasmic reticulum approaches the developing sieve plate.

The callose platelets are paired and the two members of such a pair occur on the opposite sides of a sieve plate where they are separated from each other by thin portions of the wall consisting of the middle lamella and parts of the primary walls of the two adjacent cells. The platelets increase in diameter, whereas the cellulosic bars between the pore sites decrease in width. These cellulosic bars together form the basic network of the sieve plate. The platelets and the bars increase in thickness. There is a single plasmodesma in each pore site. The perforation occurs in the centre of each platelet.

According to Evert et al. the process of pore development proceeds as follows: while the sieve-element wall increases in thickness, the callose platelets of each pair approach each other but do not yet touch near the middle lamella. At this stage a small cavity or median nodule develops around the plasmodesmatal canal in the region of the middle lamella. The canals lined with callose on either side of this nodule then widen and perforation of the pore site is thus completed. Later, callose appears also on the surface of the sieve plate. The perforation of the sieve plate takes place after the disintegration of the nucleus. Those sieve elements that have unspecialized sieve areas that are similar throughout the element are called *sieve cells*.

Sieve cells, therefore, do not contain sieve plates. These cells are usually elongated with tapering ends or their end walls are very oblique. In the positions where sieve cells overlap one another the sieve areas are more numerous. The sieve elements of the gymnosperms and vascular cryptogams are sieve cells. Elements in which sieve plates can be distinguished are called *sieve-tube members*. Sieve plates are usually found on the end walls which may be very oblique or horizontal or in intermediate planes.

In certain elements, e.g. those of *Vitis* and *Pyrus* malus, the sieve plate contains several sieve areas; while in other elements, e.g. those of *Cucurbita*, only one sieve area may be present. The former type of sieve plate is termed a *compound sieve plate* and the latter, a *simple sieve plate*. The sieve-tube members are connected one to the other by the walls that contain the sieve plates and so form *sieve tubes*. Sieve plates are found only very occasionally on the longitudinal walls of the sieve-tube members.

The walls of sieve elements are usually only primary and consist mainly of cellulose and pectin. The thickness of the wall of the sieve elements varies in different species; in some species the cell wall is 1 μ thick while in other species the wall nearly fills the cell lumen. Esau and Cheadle also found differences in the structure of the wall. In certain species they found that the wall is homogeneous, while in others the wall is composed of two layers—a thin layer close to the middle lamella and a thicker layer next to the cytoplasm. The inner layer as seen in cross-sections of fresh material has a sheen similar to mother-of-pearl, and therefore has been termed the nacreous layer. The thickness of the wall of the sieve elements usually decreases with the ageing of the element. Thick nacreous walls can be seen, for example, in *Magnolia, Laurus, Rhamnus,* and *Persea* but they are not present in *Casuarina, Crataegus, Fraxinus, Morus, Populus, Salix,* and *Passiflora,* among others.

In considerably thickened walls the inner layers are sometimes interpreted as secondary walls. The nacreous wall layer in *Cucurbita* sieve tube members was round by Deshpande to be polylammelate, the microfibrils being very densely packed and the lammellae not separating readily. The orientation of the microfibrils is very close to transverse. In *Pinus strobus* Chafe and Doohan reported that the thickened sieve cell wall comprises crossed-helical polylamellate structure in which the microfibrils are oriented in relation to the cell axis in an angle predominantly greater than 45°.

In the most primitive form the sieve elements are parenchyma cells that have undergone modifications in connection with their function. This was followed by the loss of the nucleus. This protoplasmic specialization apparently resulted in the development of the interdependence of the sieve elements and parenchyma cells that retain their nucleus, i.e. the albuminous cells in the gymnosperms and the companion cells in the angiosperms.

In the angiosperms the sieve elements and companion cells develop from the same mother cell. The specialization of the sieve elements also involved the development of thick connecting strands. In pteridophytes and gymnosperms these strands are thin and resemble plasmodesmata, while in the angiosperms they are thick and conspicuous. The evolutionary trends in the shape of the sieve elements and the arrangement of sieve areas have been more thoroughly studied in the monocotyledons.

The specialization, in this group of plants, has taken the following courses:

i) Gradual localization of highly specialized sieve areas to the end walls of the elements;

ii) Gradual changes In the position of the end wall from very oblique to horizontal;

iii) Gradual change from compound sieve plates to simple ones;

iv) Gradual reduction of the sieve areas on the side walls of the elements.

In monocotyledons the above investigators also found that the sieve tubes first developed in the aerial portions of the plants from where the development and specialization spread to the roots. This direction of development is opposite to that of the development of the vessels in the xylem. This phenomenon is understandable from a functional point of view. Features of the dicotyledonous sieve-tube members suggest that the phylogenetic development in this group of plants is similar to that of the monocotyledons, but as yet no conclusive research has been done.

Of the different types of parenchyma cells associated with the sieve-tube members the *companion cells* are the most specialized. However, the parenchyma cells are not always unequivocally distinguishable from one another even when examined in the electron microscope. The sieve-tube members and the companion cells, are related ontogenetically as they develop from the same meristematic cell. Such a meristematic cell divides longitudinally once or several times and one of the resulting cells, usually the largest, specializes to form the sieve-tube member and the others develop directly or indirectly, by further transverse or longitudinal divisions, into the companion cells. One or more companion cells may accompany a sieve-tube member.

Companion cells vary in size—they may be as long as the sieve-tube member to which they are related or they may be shorter. Companion cells may develop on various sides of the sieve tube or they may form longitudinal rows on one side only. Companion cells are strongly attached to the sieve-tube members from which they usually cannot be separated even by maceration. The walls between sieve-tube members and the companion cells are thin or possess many thin areas which, apparently, are sieve areas on the side of the sieve-tube member and primary pit fields on the side of the companion cell. The plasmodesmata on the companion-cell side are usually branched.

The length of life of the companion cells is usually the same as that of the sieve-tube member to which they are attached. This feature proves not only the ontogenetic connection but also the functional connection that exists between these cells. The companion cells and also the phloem parenchyma play an important part in the maintenance of a pressure gradient in the sieve tubes. As the companion cell differentiates, its protoplast increases in density. The mature companion cell retains its nucleus, is very rich in ribosomes, contains numerous mitochondria, rough ER, and plastids. Some companion cells contain P-protein.

In most parts of the plant the sieve elements are usually considerably wider than the companion cells. However, in the minor leaf-veins the sieve elements are often conspicuously smaller than their associated parenchyma cells. These parenchyma cells *(companion cells* or cells not ontogenetically related to the sieve elements) were called by Fischer "intermediary cells", because he believed that they mediate in the transfer of photosynthates between the mesophyll cells and sieve elements. In the small and intermediate vascular bundles of *Zea mays,* Evert did not find phloem transfer cells. He distinguished, however, two types of sieve tubes: relatively thin-walled ones that have numerous cytoplasmic connections with companion cells, and thick-walled ones that lack companion cells but possess numerous connections with vascular parenchyma cells.

Vascular parenchyma cells contain tubular extensions of the plasmalemma which enlarge to a great extent the plasmalemma surface. In gymnosperms companion cells, as described above, do not occur, but cells which stain intensely with cytoplasmic stains are present. These cells are apparently connected physiologically and morphologically to the sieve cells and have been termed *albuminous cells.* Ontogenetically these cells develop from the phloem parenchyma or from cells of the phloem rays.

However, Alosi and Alfieri found that many of the albuminuous cells in the secondary phloem of *Ephedra* are ontogenetically related to the sieve cells, as are companion cells. It was found that high respiratory and acid phosphatase activities occur in the albuminous cells associated with fully differentiated sieve cells. The increased activity is restricted to those periods during which loading and unloading of the sieve cells is taking place.

Protophloem and metaphloem

The primary phloem consists of protophloem and metaphloem. The protophloem, together with protoxylem, constitute the vascular tissue of the young elongating parts of the plant. The description of the phloem elements given above refers only to metaphloem and secondary phloem. Sieve areas cannot be distinguished in the protophloem elements of gymnosperms.

In angiosperms there are sieve-tube members in the protophloem, but in many plants there are no companion cells. These sieve-tube members are long and narrow, and sieve areas can be distinguished only with difficulty. The walls are somewhat thick and the cell contents stain only slightly. The sieve tubes of the protophloem are apparently active for a short period only. As the sieve-tube members have no nucleus they cannot divide and grow with the elongating organ, and so they become obliterated by the surrounding cells. The remnants of these obliterated cells may completely disappear in time.

In many dicotyledonous stems the parenchyma of the protophloem remains after the obliteration of the sieve-tube members, and then these cells become fibres. In leaves they form elongated collenchyma cells. In contrast to those sieve elements of the protophloem which only function for a short period and which are early obliterated, the sieve elements of the metaphloem of Pteridophyta and long-living monocotyledons, such as the Palmae, function for many years.

Formation of the cell wall and cuticle, suberization, wax deposition, and the migration of specific substances from the cytoplasm into the vacuoles represent processes of secretion. Some of these types of secretion occur in all plant cells, and some are common to certain tissues of most plants. In addition to the secretion of the above-mentioned type of materials, there are cells, groups of cells, or more complicated structures which secrete specific substances. These secretory structures occur in various plants on or in different organs and tissues.

Under the influence of studies on the processes of secretion in the animal kingdom, attempts were made to distinguish between different types of secretion in plants. Thus the terms *excretion* (elimination of end products of the metabolism) and *secretion* (elimination of substances which may still take part in the metabolic processes) were usually used.

Frey-Wyssling suggested also the use of the term *recretion* for the elimination of salts, a process which regulates the ion content of the cell. For many reasons it appears that, in plants no clear distinction can be made between secretion and excretion. Therefore one term should be used for all types of elimination of substances from the active part of the cell. In this book the term *secretion is* used. Idioblasts with various secreted substances (myrosin, tannins, mucilages, crystals, essential oils, resins, etc.) occur. Secretory idioblasts differing in size only slightly from the neighbouring cells may occur singularly or be arranged in long rows (e.g. articulated laticifers). Very large cells, sometimes as long as the whole plant (e.g. non-articulated laticifers), also occur. Secreted substances may be collected in special intercellular spaces. These spaces may be of various forms and origin. They may be more or less spherical, e.g. in *Citrus, Eucalyptus,* and termed *secretory cavities,* or they may be elongated and called *ducts* or *canals,* e.g. in the Compositae, Anacardiaceae, and Umbelliferae.

Ducts may anastomose to some extent and form two-dimensional systems according to a more or less constant pattern, e.g. in the secondary xylem of *Pinus.* Ducts may develop schisogenously, e.g. in the Umbelliferae and Pinaceae, or lysigenously, e.g. in *Copaifera* (*Caesalpiniaceae*), Mangifera (*Anacardiaceae*) and *Dipterocarpus* (*Dipterocarpaceae*). Secretory cavities and ducts may occur in various organs and tissues.

In the primary body resine ducts seem to be quite a normal feature. In the secondary body they are more liable to be influenced by external factors. In the Pinaceae it has usually been stated that in genera such as *Abies, Tsuga, Cedrus,* and *Pseudolarix* the resin ducts are produced only as a result of injury *(traumatic ducts),* while in other genera such as Pinus, *Picea, Larix,* and *Pseudot*suga they appear as a normal feature of the wood. Some workers suggest that even in the second group the ducts are much more closely correlated with injury than was thought earlier.

In the first group of Pinaceae mentioned above, the ducts are generally cyst-like and confined to tangential series at wounds. In the second group they are longer and often scattered in distribution, becoming further dispersed at a distance from the centres of injury. The more general occurrence of ducts in this second group as compared with the first one is correlated with the lengthening and scattering of ducts produced subsequently to injury rather than to normal occurrence independent of wounding as has been commonly supposed. The genus Pinus might be considered to have the ducts which are least controlled by external factors.

In normal growing trees of *Pinus halepensis* vertical ducts do not appear immediately after the commencement of cambial activity, in early spring, but only later in the season close to the formation of the late wood. Auxins were found to induce the formation of vertical resin ducts as well as an enhanced radial growth of the wood. The formation of induced vertical ducts did not start immediately after the application of auxins but only about a month later, resembling the time gap between commencement of cambial activity and duct formation in nature. Wounding and pressure also stimulate the formation of resin ducts only after a considerable amount of ductless wood has been produced. The resin duct is an elongated structure built throughout its length of cells surrounding a cavity. In the genus *Pinus* the cavity is surrounded by a layer of thin-walled and unlignified cells which are termed *epithelial cells.*

Outside these are one or more layers of cells with relatively thick, unlignified walls, which are termed *sheath cells.* The walls of these cells are apparently very rich in pectic substances. Among the sheath cells there may be some dead cells which may form a cylinder around the epithelial cells. The inner lamella of the wall of these cells consists of suberin. These cells become crushed by further expansion of the duct during growth. There is a gradual increase in the height of the duct cells from the duct cavity outwards.

The number of resin ducts in the primary body of the root was found to correspond with the number of primary xylem strands. The ducts extend from a region adjacent to the root apex, along the axis, into the hypocotyl. Some of the hypocotyl ducts may branch. The ducts end at the upper portion of the hypocotyl, at the base of the cotyledons, or they extend to the lower portions of the cotyledons.

In the region of the cotyledonary node new resin ducts develop. These ducts, which continue upwards into the cortex of the stem, have no connections with the root-hypocotyl ducts. In the seedling only juvenile leaves are present. When the seedling grows these leaves become more and more scale-like, and during the second year of the sapling brachyblasts bearing needles appear in the axils of these leaves. In the stem, each vascular bundle which diverges into a juvenile leaf passes between two cortical resin ducts. Each of these stem ducts splits periclinally into two ducts. One branch continues along the stem in the cortex and the second one enters a juvenile leaf. Thus each juvenile leaf has two longitudinal resin ducts along its abaxial side.

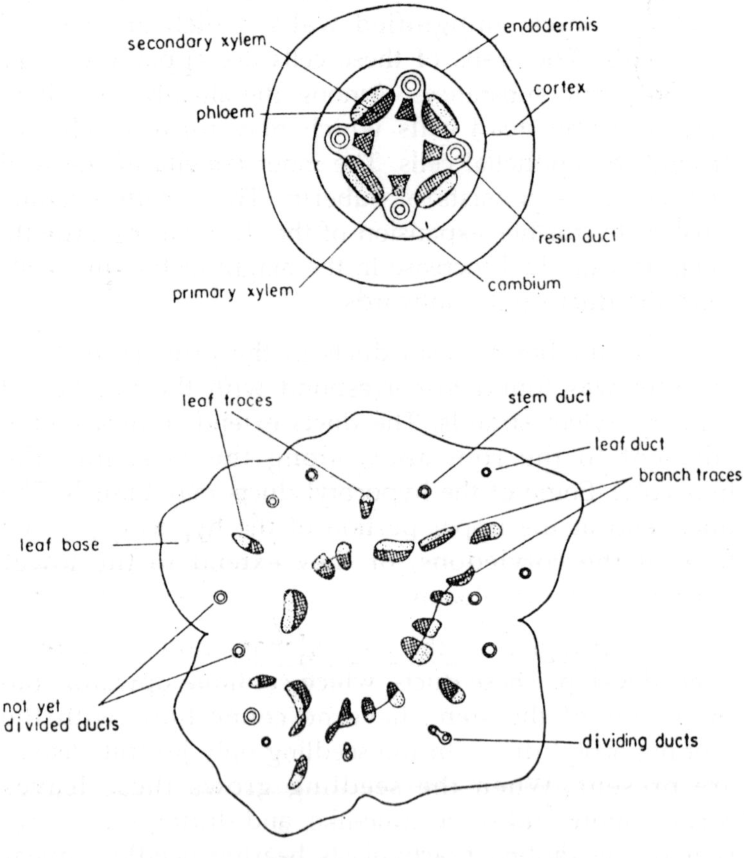

Fig. 2. (1) Cross-section of a young seedling of pinus showing a tetrarch root with four primary resin ducts. (2) a stem showing that each trace of a juvenile leaf is accompanied by two resin ducts.

A high correlation exists between the vascular and resin duct systems and both are correlated with the phyllotaxis. The same pattern prevails in the primary body of the shoot throughout the plant from a seedling to a mature tree. While the juvenile leaves change to scales no change occurs in the shoot duct system. The system becomes only more complicated by the emergence of lateral shoots. The differentiation of the stem ducts is acropetal.

In the needles of the brachyblasts, as in all other organs of *Pinus halepensis* that have secondary thickening, primary and secondary ducts are present. The primary ducts appear in the mesophyll parallel to the longitudinal axis of the leaf. In the lower portion of the leaf two lateral and 2-7 abaxial ducts can usually be distinguished. These ducts are of varying length. The secondary ducts of the needles are continuous with the ducts of the secondary xylem of the brachyblast's axis. The secondary ducts of the needle end a short distance above the needle's base.

In the secondary xylem of pines vertical and horizontal resin ducts are present. The vertical ducts of each growth ring usually appear in the outer part of the early wood and in the part of the late wood which is formed first. In the first growth-ring of tree branches, however, vertical resin ducts usually develop from the cambium immediately after the formation of 0-4 layers of tracheids.

At various levels these inner secondary ducts are tangentially split and their number increases. The inner end of each radial resin duct is connected to a vertical duct of the secondary xylem, and the lumina of the two types of ducts are continuous. Each xylary radial duct extends to the cambium, and from the cambium outwards it continues as a phloic radial duct. The outer end of each radial duct of the phloem is enlarged into a cyst-like vesicle. There are no vertical ducts in the phloem. In the cambial region the lumen of the radial duct is closed.

Radial resin ducts come in contact with vertical resin ducts Situated on the same radial plane, and connection may occasionally result between the lumina of the two types of duct. The radial ducts are found within rays which have a special fusiform shape. Connections between different vertical ducts are present only on the same radial plane, and these occur through the radial resin ducts. This means that the resin duct system cannot be considered as a spatial network. The lack of tangential connections was found to be true for all the secondary resin ducts except for those formed in the innermost portion of the first growth ring which branch and unite in a tangential direction. The vertical ducts of *Pinus halepensis* were found to be 4–10 cm in length. Apparently in the young branches the proportion of the short ducts is higher.

According to Munch, Bannan and Reid and Watson vertical ducts in old trees of other pine species may be much longer. The lack of tangential connections between the ducts of each radial plane in the secondary body and the limited length of the vertical ducts, the lack of connections between the resin systems of the primary body in the different organs and between it and the -duct system of the secondary body, may explain how differences in composition of the terpene fraction of resin could be detected at different heights of the trunk of conifer trees and between different organs and tissues of the plant.

A special type of duct, termed a *kino vein,* occurs in the wood of the genus *Eucalyptus*. Kino differs from gum in containing polyphenols. The kino veins vary considerably in size: from a few isolated veins to a dense anastomosing mass completely encircling a tree, and 1.5 to 5 mm in width. The length of the veins varies from a few centimetres to 3 M.

Kino veins form in the cambial region as a result of injury. When a patch of veins of moderate size is formed, as seen in tangential view, the veins extend for a greater distance upwards from the wound than downwards. Kino veins develop in the zone of traumatic parenchyma, which is produced by the cambium shortly after the stimulus to vein formation is given and in which groups of cells accumulate large quantities of polyphenols.

At particular foci, groups of cells containing polyphenols break down and form ducts into which the contents of the kino-producing cells are released. At about the same time the cells surrounding the future veins divide and form a peripheral "cambium". The derivatives of this "cambium" grow, accumulate polyphenols, break down, and enlarge the quantity of kino already present in the ducts. In the final stage the peripheral cambium produces a layer of derivatives which becomes suberized in the form of a typical periderm.

Laticifers

Laticifers are present in a large number of species and genera belonging to about twenty families. These belong mostly to the Dicotyledons, but laticifers were also found in a few monocotyledonous families and in the genus *Regnellidium* of the Marsiliaceae (Pteridophyta).

Latex, a suspension or, in certain, cases an emulsion, of many small particles in a liquid with a varying refractive index, is known to occur in many angiosperms. This liquid may occur either in series of cells or in single long cells; both these specialized structures have been termed *laticifers.*

Secretory cells occurring in the conifers are also regarded as laticifers. The fact that laticifers are restricted to a small number of families, between many of which there is no evidence of close taxonomic relationship, suggests that the capacity to produce latex has evolved more than once. The chemical composition of the suspended or emulsified particles of the latex differs in the different species.

Among the suspended material rubber particles [C_5H_8)n], waxes, resins, proteins, essential oils, mucilages, and in certain *Euphorbia* species, variously shaped starch grains can be found. Latex, like the cell sap, also contains salts, organic acids, and other substances in true solution. Certain plants contain, in the latex, sugars (the Compositae), tannins *(Musa),* alkaloids *(Papaver somniferum),* and in *Carica papaya,* a proteolytic enzyme, papain, is present.

The colour of latex varies in the different plant species—it may be white and milky *(Euphorbia, Lactuca, Asclepias),* yellow-brown *(Cannabis),* yellow to orange *(Chelidonium)* or colourless *(Morus).* The function of the latex in the plant is not clear. Haberlandt believed the latex to be of nutritional value in many plants. Sperlich regarded latex as a reserve material.

The most acceptable theory is that the latex contains metabolic by-products and the laticifers represent a secretory system. Substances that are regarded as end products rather than as reserve materials accumulate in the laticifers. Rubber, for example, like the essential oils and unlike starch, cannot revert to nutritional substrates as no enzymes capable of breaking down rubber are present in the plant. In the laticifers latex is under pressure and therefore when the laticifers are cut the latex is extruded. This is a pressure flow. The dynamics of latex flow have been dealt with in detail by Frey-

Wyssling. Heinrich, who studied the laticifers of *Taraxacum bicorne* and Ficus *elastica* with aid of the electron microscope, expressed the view that rubber particles are synthesized in the cytoplas and not in the plastids as thought before. In *Ficus* the rubber particles were surrounded by an envelope, possibly of membranous nature. As seen in *Taraxacum,* the particles remain in the cytoplasm and only in old laticifers, when the tonoplast disintegrates, are they freed. Many vacuoles appear in the initials of the laticifers. Thureson-Klein, who studied the laticifers of *Papaver somniferum,* has noticed in the cytoplasm the appearance of many vesicles with dense material which may contain alkaloids.

In very young laticifers of *Euphorbia marginata,* Schulze et al. found rubber drops first in the cytoliasm, but later almost only in vacuoles. In later developmental stages a large central vacuole bounded by a tonoplast is formed. The small vacuoles with rubber particles join the central vacuole which becomes more and more filled with rubber. A tonoplast was also reported by Schnepf to occur in mature laticifers of *Euphorbia pulcherrima.* Schulze *et al.* concluded that in the mature articulated-branched laticifers of *Taraxacum, Papaver,* and *Hevea* the rubber particles appear as single drops in the cytoplasm, whereas in the non-articulated laticifers of *Euphorbia* they appear assembled in central vacuoles.

Studies of *Ficus* and *Taraxacum* showed that in the cytoplasm of developing laticifers a relatively well-developed endoplasmic reticulum and many ribosomes occur. The mitochondria are small and degenerate when the rubber particles start to appear. Golgi bodies appear only in the growing portions of the laticifers. In old laticifers the disintegration of the organelles, as seen in *Taraxacum,* proceeds in the following sequence: Golgi bodies, ribosomes and mitochondria, the ER, the nucleus, and plastids. In *Ficus* mitotic activity of the nuclei seems to be confined to apical zones. In the exuded latex only degenerated nuclei and plastids, and more rarely mitochondria, appear. In mature laticifers of *Ficus,* filaments similar to those in sieve elements were observed.

The cell wall of the laticifers is entirely primary and may be as thick as or thicker than the neighbouring parenchyma cells. The thick walls contain cellulose and a high proportion of pectic substances and hemicelluloses. These walls are highly hydrated. Both the thick walls and the thin ones, which do not differ from the walls of the neighbouring parenchyma cells, are very elastic. There is controversy as to the occurrence of primary pith fields.

In the laticifers of some plants the protoplasts are reported to remain intact even after maturation; in others the cytoplasm, including the organelles, may undergo degeneration. In some plants the mature laticifers were reported to contain large central vacuoles bounded by a tonoplast, e.g. in *Euphorbia marginata*. In *Taraxacum bicorne* the tonoplast disappears. In addition to the usual components of the cytoplasm, electron-dense globules and small vacuoles or vesicles occur in laticifers.

In the laticifer initials of *Papaver somniferum* the ER was found to be abundant. In the initials of the *Taraxacum bicorne* laticifers Golgi bodies were observed to be rare and to degenerate before the appearance of the first rubber particle. The relatively numerous mitochondria remain small and their inner structures undergo early lysis. The degeneration of the mitochondria and ribosomes occurs long before the dissolution of the end wall. Because of their distribution in the plant body and their liquid content that flows out promptly when the plant is cut, the early botanists called them *Lebenssaftgefasse* and compared them with the blood vessels of animals. Their association with the vascular bundles, especially with the phloem, brought the researchers later to the assumption that the laticifers take part in translocation of assimilates. This however, is incorrect.

The laticifers at the separation layer of leaves of *Hevea* which have completed their development were found to be occluded by callose. This would prevent translocation of substances from leaves to the stem via laticifers. The advantage of association with phloem may be an efficient supply of initial latex precursors from the assimilating tissues.

The laticifers have also been thought to take part in storage of food materials. The results of experiments contradicted this assumption too. The food materials present in the latex are not readily mobilized when the plant is grown under conditions unfavourable for carbohydrate synthesis. In most cases the laticifers apparently produce and store mainly substances which do not re-enter the plant metabolism. The most probable main function of the laticifers seems to be protection. The latex may play a role in covering wounds, as a defence against herbivores, and perhaps also as a defence against microorganisms.

The laticifers are divided into two main types: *non-articulated* and *articulated*. This classification has no relationship to taxonomic groups and thus different types of laticifers may be found in different species of one family. The non-articulated laticifers develop from a single cell which greatly elongates with the growth of the plant and which is sometimes branched. Such laticifers are also termed *laticiferous cells*.

Articulated laticifers consist of simple or branched series of cells which are usually elongated. The end walls of such cells remain entire or become porous or disappear completely. Such laticifers are also termed *laticiferous vessels.*

Non-articulated laticifers are characteristic of various species of the following families: Apocynaceae, Asclepiadaceae, Euphorbiaceae, Moraceae, and Urticaceae. Simple (unbranched) nonarticulated laticifers are found, for example, in *Vinca, Urtica,* and *Cannabis,* and branched nonarticulated laticifers in *Euphorbia. Nerium,* Ficus, and *Asclepias.* As has already been mentioned, different forms of non-articulated laticifers exist and in certain mature plants laticiferous cells may develop into very large systems which extend throughout the shoot and root tissues. Scharffstein Rosowski who worked on some *Euphorbia* species express the view that their entire laticifer system is derived from a few initials that are already present in the embryo. Mahlberg found in *Nerium* that the number of initials is constant and that they can all be distinguished already in the embryo where they appear in the cotyledonary node from where they send branches into the cotyledons, the hypocotyl and the radicle.

In the course of development of certain species of *Euphorbia* laticiferous cells are found on the circumference of the central cylinder and they branch into the leaves and pith. In the leaves of certain species of the Euphorbiaceae the laticiferous cells reach the epidermis where they may even come into contact with the cuticle. Blaser working on *Cryptostegia, grandiflora* (Asclepiadaceae) found that the early formed laticiferous cells in the cortex branch 'radially in the position of the leaf-gaps and penetrate into the pith. After a period of cambial activity these branches of the laticiferous cells become surrounded by the secondary phloem and xylem. Laticifers were observed in the rays of the secondary xylem of some plants.

In a number of species of *Artrocarpus* (Moraceae) non-articulated laticifers branch in the wood both in the radial and in the axial directions. The growth and expansion of the branched laticiferous cells is continuous throughout the life of the plant and their ends penetrate into the new buds, developing leaves and- growing regions of the root. Such development of laticiferous cells has been observed, for example, in *Nerium*.

The growth of these branched laticiferous cells is by a combination of intrusive and symplastic growth. It is of interest that in certain plants even the old portions of these cell walls retain the ability to grow and produce new branches, as was by grafting experiments carried out by Scharffstein on *Euphorbia esculenta.* Numerous nuclei have been observed in the branched nonarticulated laticifers of different plants. Scharffstein and Mahlberg drew attention to the fact that the multiplication of the nuclei takes place early in the ontogeny of the laticiferous cell.

The growth of the simple non-articulated laticifers is simpler than that of the branched nonarticulated ones. The initials of the former have not been observed in the embryo but only in the developing shoot, as, for instance, in *Vinca* and *Cannabis*. The initials appear below the apical meristem and they develop into long, unbranched ducts which grow by a combination of intrusive and symplastic growth. These laticiferous cells may penetrate from the stem into the leaves *(Vinca)* or they may develop independently in these organs *(Cannabis)*.

Articulated laticifers are characteristic of different species of the Compositae, Convolvulaccae, *Papaveraceae, Euphorbiaceae, Caricaceae, Sapotaceae, Liliaceae*, and *Musaceae*. Simple non-anastomosing articulated laticifers occur in *Musa, Allium, Convolvulus,* and *Ipomoea*. Anastomosing articulated laticifers are found in *Sonchus*,

Cichorium, Lactuca, Taraxacum, Tragopogon, Scorzonera, Carica, Manihot, Papaver, and *Hevea.*

Like the non-articulated laticifers, the articulated laticifers appear in the early ontogenetic stages. In *Taraxacum kok-saghyz* the differentiation of the laticifers commences with the uptake of water at the time of germination of the seed. The primary laticifers appear in the pericycle in close connection with the phloem, whereas the secondary laticifers differentiate in the secondary phloem very close to the cambium. Scharffstein, Sperlich, Artschwager and McGuire, and other workers observed the process of the formation of the articulated laticifers from single cells in the embryos of *Tragopogon, Scorzonera, Taraxacum,* and *Hevea brasiliensis.*

Articulated laticiferous vessels develop in many plants in the phloem or pericycle of the stem and root. They also occur in the leaf mesophyll. Apart from the connections formed between the cells of one laticifer by the entire or partial dissolution of the end walls, connection may also be made between adjacent laticiferous vessels in species with branched articulated laticifers.

Articulated laticifers may also become connected by horizontally or diagonally orientated elements. Reticulate systems may also develop as a result of the formation of branches which arise on the side walls of the laticifers. Sometimes parenchyma cells between two laticiferous vessels redifferentiate and so connect the two vessels. The laticifer system in the capsules of *Papaver somniferum* was studied by several workers. The laticifers in this plant differentiate soon after seed germination. They appear in all organs but they are most abundant in the ovary. Here they produce a peripheral-network the main strands of which are parallel to the vascular bundles and are interconnected by anastomoses. The network becomes denser with the development of the capsule. The laticifers of *Hevea brasiliensis* appear mainly in the bark, but they occur also in the pith and leaves.

The most important laticifers as a source of rubber are those which appear in the inner bark. These are derived from the cambium. The laticifers are arranged in concentric zones, with lateral tangential anastomoses. Radial connections apparently occur only as a result of disturbances in the cambial activity. The number of rings of laticifers, the density of the network of each zone and the diameter of the individual laticifers vary from one clone to another. Correlation between the various anatomical features of the laticifers and rubber yield is restricted to clones and cannot be drawn for populations.

Grafting experiments showed that laticifers become united. Bouychou found that laticifers in *Hevea* were produced from the cambium of tissue grown *in vitro* and the segments of laticifers were still capable of producing rubber. The simple (unbranched) laticifers of *Musa* usually accompany the vascular tissues. In Musa the central portion of the end walls of two adjacent cells torn away, but usually it remains to one side of the cell where it appears as a loose flap. In *Allium* the laticiferous vessels are in no way connected with the vascular tissue and they are found in the third layer of leaf mesophyll or in the third layer below the abaxial epidermis of the bulb scales.

Epidermis

Morphologically the epidermal cells are not uniform and among them, apart from the ordinary cells, many types of hairs, stomatal guard cells, and other specialized cells are found. Topographically, however, and to a certain extent also ontogenetically, the epidermis constitutes a uniform tissue. The earliest stages of the ontogenetic development of the epidermis differ in the root and shoot. This fact caused certain investigators to coin special terms, *epiblem* and *rhizodermis,* for the outermost layer of the root. However, if the development of the epidermis from the protoderm is traced ignoring the problem of the origin of this meristematic tissue, it is possible to apply the term *epidermis to* all the organs of the various groups of vascular plants.

The epidermis usually exists throughout the entire life of those organs that have no secondary thickening. In a few plants, such as long-lived monocotyledons with no secondary thickening, the epidermis is replaced by a cork tissue as the organs age, The duration of the epidermis in organs with secondary growth differs; usually in stems and roots the epidermis is replaced, by the periderm during the plant's first year, but there are certain trees, e.g. *Acer striatum,* in which the periderm develops only after several years of secondary growth of the organ. In such, cases the epidermal cells continue to divide anticlinally and to enlarge tangentially.

Various types of epidermal cells can be distinguished in different plants: the ordinary cells of the epidermis; single cells or groups of cells with special structure, form or content; cells connected with stomata; and epidermal appendages termed *trichomes.* The ordinary cells of the epidermis vary in shape, size, and arrangement, but they are always closely attached to form a compact layer devoid of intercellular spaces. In the epidermis of petals, air spaces may sometimes occur but they are always covered by the cuticle. Many epidermal cells are tubular, and, in the leaf blade of dicotyledons the anticlinal walls are mostly sinuous. In the stems, and especially in the leaves of many monocotyledons, the epidermal cells are elongated. In the epidermis of certain seeds (in species of the Leguminosae and in *Punica)* the cells are relatively very elongated in a radial direction and are rod-shaped. In certain plants, for instance *Aloe aristata,* the epidermal cells appear to be hexagonal in surface view but actually they are polyhedral, and according to Matzke, the average number of faces is 10.885.

The external wall of the epidermal cells of certain leaves and petals is raised in the form of papillae. Haberlandt believed that the papillae may have a function in concentrating light which is limited for plants growing in the shade. In certain pteridophytes such papillae are found on the epidermal cell wall facing the mesophyll.

The epidermal cell wall differs in thickness—some cells are thin-walled while in others the outer or the outer and inner periclinal walls are thicker than the anticlinal walls. In seeds, scales, and certain foliage leaves, e.g. the leaves of conifers, the epidermal cell walls are very thick and lignified. Outside the outer cellulose wall there is usually a layer of more or less pure pectin. This layer enables the isolation, of the cuticle from many leaves with the aid of pectinases or by other means. In many secretory trichomes the cuticle becomes detached during the process of secretion.

In leaf petioles and epicotyles of some plants in which the periclinal walls are thicker than the anticlinal walls, the periclinal walls were found to consist of lamellae in which the cellulose microfibrills are oriented alternately, parallel or transverse to the longitudinal cell axis. This arrangement of fibrils resembles that of collenchyma cell walls mentioned above. Primary pit fields and plasmodesmata are often found in the walls, especially on the radial and inner walls of the epidermal cells. In epidermal cells of the aerial organs there occur bundles of specific interfibrillar spaces, mostly in their outer walk. These bundles of spaces in the cellulosic wall obtained the name *ectodesmata, and* can be demonstrated by specific reagents. In cross-section of the wall they form thread-shaped, ribbon-like or cone-like structures. Occasionally mushroom-like forms, the broader part of which faces the cell lumen, also occur.

It has recently been shown that ectodesmata in contrast to plasmodesmata are not plasmatical structures. A suggestion was made that they may serve as pathways of penetration of solutions to and from the protoplast. In order to, distinguish sharply between ectodesmata and plasmodesmata Franke has proposed to use the term *teichodes (teichos* = wall, hodos = path) instead of ectodesmata. In the epidermal cells of certain leaves, petals, and dry fruits, partial partitions are found on the inside of the outer wall. These partitions sometimes almost reach the inner epidermal wall.

Cutin, a fatty substance, is usually present in the outer walls of the epidermal cells. This substance is found within the cell wall, i.e. in the interfibrillar and intermicellar spaces of the cellulose, and it also constitutes a special layer—the *cuticle*—on the outer surface of the cell wall. Cutin stains red with Sudan IV.

All the parts of the herbaceous stem, the leaves, and, to a certain extent, the mature portions of the root, are covered with a cuticle. The cuticle is usually absent from the actively growing parts of the roots. The cuticle possesses in some genera (e.g. *Eucalyptus*) a complex and species-specific ornamentation which can be used for diagnostic purposes. In certain plants, which have been preserved from early geological eras, the cuticle has retained its shape, and the structure of the epidermal cells can be learnt from it. These preserved fragments of cuticle are often used to classify these early plants. The cuticle is of varying thickness in different plants, and it is usually thicker in plants growing in dry habitats. The surface of the cuticle may be smooth, rough, ridged, or furrowed. Priestley states that very thin layers of cutin can be distinguished on those mesophyll cell walls that border the intercellular spaces. These spaces form a continuous system connected with the stomata, and therefore these cutin layers constitute a continuation of the cuticle on the outer surface of the epidermis.

The cuticle on the surface of the epidermis often penetrates, to a certain extent, between the radial walls of the epidermal cells. The cuticle consists of two layers; an outer layer built of cutin only—the cuticle proper, and a layer below it—the *cuticular layer* which consists of cutin and wall materials. The cuticle proper is formed by the secretion of cutin or its precursors to the surface of the cell wall. This process is called *cuticularization.*

The cuticular layer is formed by deposition of cutin between the cellulose microfibrils of the outermost wall layers, where pectin and hemicellulose may also be present. This process of cutin deposition is called cutinization. Between the outer periclinal cellulose wall and the cuticular layer there is in many plants a layer very rich in pectin—the *pectin layer.* This layer is believed to be continuous with the middle lamella of the anticlinal walls of the epidermal cells.

In the cuticular layer birefringent wax may be embedded. In the stems of the xerophyte *Monetta aphylla* this layer may reach a thickness of 140-180 μm. Electron microscopical studies suggested that the cuticle appears first as a thin, lamellate layer (the cuticle proper). Later another cutin layer containing a reticulum of electron dense material, apparently polyuronides becomes visible (the cuticular layer). This layer eventually becomes the largest part of the cuticle. In *Hakea suaveolens* the reticulate cuticular layer was found to develop long before the final expansion of the cells and parallel with the late phase of lamellae differentiation. Deposits of *wax* in the form of granules, as in *Brassica, Dianthus,* or of rods, as in *Saccharum* or as continuous layers, as in *Thuja orientalis,* are often found on the surface of the cuticle. Wax may also be deposited in the form of platelets or scales. The wax gives the "bloom" of many leaves and fruits, and is important in reducing the wettability of surfaces. Some wax mixtures fail to crystallize and may form an oily layer of irregular flat plates above the cuticle as, for instance, in many apple fruit surfaces.

As seen under the scanning microscope, waxes in crystalline state may form very elaborate structures. In the wax palm *(Copernicia cerifera)* and in *Ceroxylon andicola* the wax layer may be 5 mm thick. The structure and amount of the surface wax has an extremely important effect on the degree to which a surface can be wetted. This phenomenon represents one of the factors which govern the selective effect of herbicides on some crop plants. In the leaves of *Agave* a continuous layer of wax is also found below the cuticle. In certain plants platelets of wax have been found within the cutin of outer epidermal cell walls.

Cuticles develop during the early stages of growth of the organs. The precursor of cutin migrates in the form of minute droplets through the matrix of the epidermal wall. The precursors seem to be unsaturated fatty acids. The cuticle is built up centripetally, its outer zone being formed first. The cuticular material hardens gradually, due to a continuous process of oxidation and polymerization. Various theories have been proposed to explain this outward movement of the cutin.

According to some investigators, special channels are present in the outer epidermal walls through which precursors of cutin pass; other investigators believe that the porous nature of the wall suffices. The way in which the surface wax is extruded is still disputed. Some workers suggest that the exodesmata or special canals in the wall and cuticle are involved in this process. Other workers, on examination of a wide range of plant cuticles, have not detected canals. Schieferstein and Loomis doubt the existence of true canals. They suggest that the wax is extruded only through the fragile cuticle of young leaves and that later extrusion is prevented by the thickening of the cuticular layer.

Deposits of salts in the form of crystal, e.g. in *Tamarix* and *Plumbago capensis*, of caoutchouc, e.g. *Eucalyptus,* or of oils and resins, sometimes occur on the surface of cuticle or within it. Deposits of silicon salts are found in the epidermal cell walls of many plants, as, for example, *Equisetum,* the Gramineae, many species of the Cyperaceae, the Palmae and certain species of the Moraceae, the Aristolochiaceae, and the Magnoliaceae. Lignin is rarely found in the epidermal cell walls.

When it is present it may be found in all the walls or only in the outer wall. Lignified epidermal walls are found in the leaves of the Cycadaceae, in the needles of conifers, in the rhizomes of the Gramineae, in the strips of epidermis above the bundles of sclerenchyma in the leaves of the Gramineae, Juncaceae, and Cyperaceae, in the leaves of certain species of *Eucalyptus* and *Quercus*, and in *Laurus nobilis* and *Nerium oleander.*

Parts of the epidermal cell walls of groups of cells or of single cells may become mucilaginous in certain dicotyledonous families, such as the Moraceae, Malvaceae, Rhamnaceae, Thymelaeaceae, and Euphorbiaceae. In certain seeds, such as those of *Linum usitatissimum* and species of *Alyssum,* the outer walls of the epidermal cells become mucilaginous. In the nectaries of certain plants the epidermal cells also become mucilaginous at the time of nectar secretion.

The epidermal cells are usually highly vacuolated. However, those of *Eucalyptus papuana* leaves were observed to have only few small vacuoles. It is generally assumed that the epidermal cells contain leucoplasts, but these could not be seen in several species studied, e.g. species of *Eucalyptus, Ligustrum,* and *Phaseolus*. Chloroplasts are found in certain pteridophytes, hydrophytes, and in some shade plants.

Numerous mitochondria, ER, spherosomes, and Golgi bodies are usually present. Anthocyanins are found in the vacuoles of the epidermal cells of the petals of numerous flowers, in the leaves of *Zebrina pendula* and red cabbage, in the stems and petioles of *Ricinus,* and in different organs of many other plants. Tannins, mucilage, and crystals may be present in epidermal cells. In certain pteridophytes, in the gymnosperms, in many species of the Gramineae, and certain dicotyledons, fibre-like epidermal cells are found.

In the Gramineae, between the elongated epidermal cells, i.e. *long cells,* above the veins, there are *short cells* which are of two types—silica cells and cork *cells.* The latter two types of cells often occur successively in pairs, throughout the length of the leaf. The fully developed silica cells contain silica bodies which are isotropic masses of silica in the centre of which are usually minute granules. In surface view the silica bodies may be circular, elliptic, dumb-bell, or saddle-shaped. The silicon was reported to be present, if at all, only in small amounts in the young silica cells; its accumulation being rapid in senescing cells. The walls of the cork cells are impregnated with suberin and many of them contain solid organic substances. The above short cells sometimes bear papillae, setae, spines, or hairs.

Metcalfe draws attention to the fact that the cork cells in many plants contain silica bodies, and that in certain grasses silica bodies also occur in some elongated cells. Silica bodies also occur in specialized epidermal cells of the Cyperaceae and some other monocotyledons. In the Gramineae and many other monocotyledons, with the exception of the Helobiae, *bulliform cells* are found in the epidermis. These cells are larger than the typical epidermal cells and they are thin-walled and have a large vacuole. The bulliform cells may constitute the entire adaxial epidermis of the leaf or they form isolated parallel strips in the area between the veins.

In a cross-section of the leaf these cells appear in a fan-like arrangement in which the central cell is the tallest. In certain plants bulliform cells are also found on the abaxial surface of the leaf. These cells are sometimes accompanied by similar mesophyll cells. Bulliform cells contain much water and are devoid, or nearly so, of chloroplasts. According to one opinion they function in the opening of the rolled leaf as present in the bud. According to a second view they bring about the rolling or unrolling of mature leaves as a result of their loss or uptake of water.

An investigation carried out by shields on twelve xerophytic grass species suggests that, in the opening of the young leaves from the bud and in the hygrochastic movements of the mature leaves, elements other than the bulliform cells may actively participate. According to Metcalfe, the bulliform cells often become filled with large masses of silica and their outer walls often become thick and cutinized. Other specialized epidermal-cells are the lithocysts, which are found in the Acanthaceae, Moraceae, Urticaceae, and Cucurbitaceae.

In the Cruciferae *myrosin cells* are sometimes found in the epidermis. These cells are large secretory cells which contain the enzyme myrosin, and they stain red in the Millon test, or violet with orcein solution and concentrated hydrochloric acid. Large idioblastic cells containing mucilage occur in the epidermis of Lythraceae and very large sac-like cells which intrude into the mesophyll were observed in *Cleome aspera*. Secretory cavities may sometimes develop from epidermal cells, e.g. in *Psoralea*. Long pipe-like cells filled with tannic substances appear in the epidermis of *some Saxifraga* species.

Stomata

Continuity of the epidermis is interrupted by minute openings. These are intercellular spaces each of which is limited by two specialized cells termed the *guard cells*. The guard cells, together with the opening between them, constitute the stoma. In many plants *subsidiary* or accessory cells can be distinguished. These cells differ morphologically from the typical epidermal cells and they constitute two or more cells bordering the guard cells to which they are apparently functionally connected. The stoma, together with the subsidiary cells if present, is termed the stomatal apparatus or the stomatal complex. The subsidiary cells usually develop from protodermal cells adjacent to the stomatal mother cells, but they may also develop from sister cells of the mother cell. Based on the ontogenetic relationship between the guard and subsidiary cells, stomata can be divided into three types:

i) *Mesogenous stomata,* in which the subsidiary cells have a common origin with the guard cells,

ii) *Perigenous stomata,* in which the subsidiary cells develop from protodermal cells adjacent to the stomatal mother cell; and

iii) *Mesoperigenous stomata,* in which the cells surrounding the stoma are of dual origin, one or sometimes more of the subsidiary cells having a common origin with a guard cell, while the other or others have not.

An ontogenetic system of classification of stomata based primarily on the mode of division of the guard mother cell with respect to orientation of the wall which cuts it from the meristemoid has recently been suggested by Payne.

Stomata are usually found on the aerial portions of the plant and especially on leaves, ordinary stems, and rhizomes. Stomata are absent from roots and the entire plant body of certain parasitic plants that lack chlorophyll, such as *Monotropa* and *Neottia.* In *Orobanche,* however, stomata are found on the stem although this genus is also devoid of chlorophyll.

Stomata are present in some submerged water plants but not in most others. Stomata may be found on petals, staminal filaments (e.g. *Colchicum),* carpels, and seeds, but these stomata are usually non-functional. In photosynthesizing leaves stomata may be found on both sides of the leaf or on the lower side only. In certain water plants with leaves that float on the surface of the water, e.g. *Nymphaea,* stomata are found only on the upper surface of the leaf which is exposed to the atmosphere. In leaves with reticulate venation the stomata are distributed in no particular order, while in leaves in which the majority of veins are parallel, as in the Gramineae, the stomata are arranged in parallel rows.

The guard cells of the stomata may be level, sunken, or raised relative to the other epidermal cells. In cases where the epidermis is multiseriate, e.g. as in species of *Anabasis* and *Haloxylon* and in Ficus *elastica,* the guard-cell mother cells are differentiated at that stage of development when the protoderm is as yet a single layer. During the process of further development the surrounding protodermal cells undergo several periclinal divisions resulting in the raising of the multiseriate epidermis above the level of the guard cells.

Below the stomata and directed inwards to the mesophyll are large intercellular spaces which are termed *substomatal chambers*. The guard cells of most plants, except those of the Gramineae, Cyperaceae, and some others, are kidney-shaped in outline. The size of the aperture between the guard cells increases or decreases as a result of turgor changes in the guard cells. In *Vicia faba* it was found that the average guard cell volumes were 4.8×10^{-12} litres per stomatal apparatus in open stomata and 2.6×10^{-12} litres for the ones in closed stomata. As a result of unequal thickness and consistency of the walls, the guard cells undergo changes in form with the increase in volume. These changes cause the opening of the stoma. In most cases the thinnest wall is that closest to the subsidiary cells, and is termed the *back* or *dorsal wall.* With the increase of turgor this thin, flexible back wall expands more than the other walls, which are more resistant to pressure. As a result the shape of the guard cells changes in such a way that the stomatal pore opens.

According to another hypothesis the radial arrangement of the cellulose microfibrils (radial micellation) in the guard cell walls plays a major role in stomatal movement. In guard cells with very thick periclinal walls, e.g. in *Araucaria, Agathis,* and *Anabasis articulata,* the additional wall layers consist of numerous lamellae in which the cellulose microfibrils alternate in their orientation. Protuberances of the guard-cell wall may be present above, or both above and below, the stomatal aperture. In cross-section these protuberances appear as horn-shaped ledges. The outer ledge delimits the front cavity above the aperture, and the inner ledge delimits the back cavity Which abuts on the substomatal chamber. In some Eucalyptus species the cuticle forms special ridges termed stomatal bars.

The guard cells in the Gramineae and Cyperaceae are different in form from that described above. They are elongated and dumb-bell shaped. The ends of these guard cells are expanded and thin-walled, while the middle portions are elongated and thick-walled and the cell lumen is narrow. As a result of turgor increase in this type of guard cell, the expanded ends swell and so push apart the middle elongated portions of the cells. Because of the above structure the nuclei in the guard cells of grasses appear as two elipses connected by a narrow thread. According to Flint and Moreland the two parts of the nucleus may become completely separated. In the stomata of the Gramineae, as seen with the aid of the electron microscope, the wall between the ends of the guard cells is incomplete and the protoplasts are confluent.

The chemical composition of the guard cell wall is the same as that of the ordinary epidermal cells of the same plant. They are usually covered by cuticle which generally continues on that wall that faces the aperture and which also reaches the cells abutting on the substomatal chamber. In citrus, cuticle is absent from the cell wall facing the stomatal aperture. Apart from the types of guard-cell structure described above many other structural variations exist in the mono-and dicotyledons, e.g. in species of *Haloxylon* and *Anabasis,* in the Palmae.

Guard cells of many vascular cryptogams, gymnosperms and some angiosperms possess lignified wall thickenings. Kaufman suggested that the lignified wall areas are related to the opening mechanism of the stomata. Electron microscope studies have shown that the guard cells contain numerous mitochondria, elements of ER, Golgi bodies, and vacuoles of various sizes. However, in some investigated species of *Paphiopedilum* (Orchidaceae) internal lamellar structures and chlorophyll are completely lacking in the guard-cell plastids although the stomata are functional. Starch grains occur in the plastids. Complete plasmodesmatal connections between guard cells and subsidiary cells were not observed to occur in mature stomata of most plants investigated with the aid of the electron microscope. In developing stomatal complexes, plasmodesmata were found between guard cells and subsidiary cells and also between sister guard cells.

In mature stomata of *Phaseolus vulgaris* Willmer and Sexton reported that aborted incomplete plasmodesmata occasionally occur in the common guard-subsidiary cell wall. The plasmalemma shows numerous invaginations.

According to the relation of stomata to the neighbouring epidermal cells, the stomata have been classified into various types, This classification is separate from the one based on development. Although different types may occur in one and the same family, or even in a leaf of the same species. The structure of the stomatal apparatus can be of use in taxonomic studies.

Morphologically, four main types of stomata have been distinguished in the dicotyledons on the basis of the arrangement of the epidermal cells neighbouring the guard cells.

i) *Anomocytic type* in which the guard cells are surrounded by a certain number of cells that do not differ in size and shape from the other epidermal cells. This type is common in the Ranunculaceae, Geraniaceae, Capparidaceae, Cucurbitaceae, Malvaceae, Scrophu-lariaceae, Tamaricaceae, and Papaveraceae.

ii) *Anisocytic type* (cruciferous) in which the guard cells are surrounded by three unequally sized subsidiary cells. This type is common in the Cruciferae, in *Nicotiana, Solarium, Sedum,* and others.

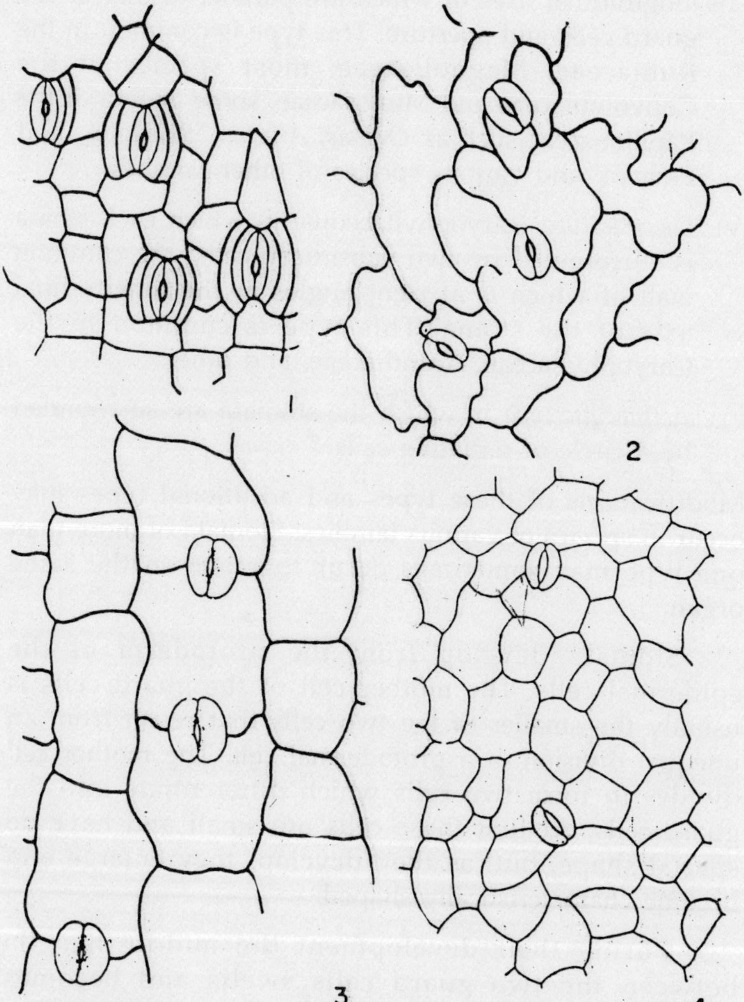

Fig. 3. Variuos patterns of stomatal arrangements in leaves of variuos plants. (1) paracytic type, (2) anisonic type, (3) diacytic type, (4) anomocytic type.

iii) *Paracytic type* (rubiaceous) in which each guard cell is accompanied by one or more subsidiary cells, the longitudinal axes of which are parallel to that of the guard cells and aperture. This type is common in the Rubiaceae, Magnoliaceae, most species of the Convolvulaceae and Mimosaceae, some genera of the Papilionaceae such as *Ononis, Arachis, 'Phaseolus,* and *Psoralea,* and various species of other families.

iv) *Diacytic type* (caryophyllaceous) in which each stoma is surrounded by two subsidiary cells, the common wall of which is at right angles to the longitudinal axis of the stoma. This type is common in the Caryophyllaceae, Acanthaceae, and others.

v) *Actinocytic type* in which the stomata are surrounded by a circle of radiating cells.

Modifications of these types and additional types may occur in species of various families. Stomata of more than one type may sometimes occur together on the same organ.

Stomata develop from the protoderm of the epidermal cells. The mother cell of the guard cells is usually the smaller of the two cells that result from an unequal division of a protodermal cell. The mother cell divides to form two cells which differentiate into the guard-cells. At first these cells are small and have no special shape, but, as they develop, they enlarge and become characteristically shaped.

During their development the middle lamella between the two guard cells swells and becomes lens-shaped shortly before the time when it disintegrates to form the stomatal aperture. Stevens and Martin suggested that, whilst pore formation is initiated enzymatically, the separation of the two guard cells is brought about by osmotic force derived from starch hydrolysis. Even in those cases in which the mature guard cells are sunken or raised relative to the ordinary cells of the uniseriate epidermis, the guard cell mother cells and the guard cells are level with the other epidermal cells immediately after their formation. The sinking and the raising is brought about during the maturation of the guard cells.

The development of stomata in the leaf continues for a relatively long period during the growth of the leaf. Arrested stomatal development and abnormal stomata with a single guard cell have been reported. The first step in the differentiation process is the asymmetric division of certain protodermal cells. Before the cell divides, the nucleus becomes displaced in one direction and the vacuoles occupy the other end of the cell. The nucleus then divides. One of the daughter nuclei which is closer to the vacuoles becomes larger and more weakly staining than the other at the denser end of the dividing cell. The smaller distal nucleus divides later to form the two guard cells, but before doing so it induces division of neighbouring epidermal cells. In these cells also there is denser cytoplasm adjacent to the guard mother cell, and the induced divisions are also asymmetrical.

Prior to each of the asymmetric divisions a band of microtubules appears in the peripheral cytoplasm. This band indicates the location on the wall of the dividing cell where the cell plate will join it at the final stage of division. Thus the future plane of cell division is indicated by these microtubules at preprophase.

On the basis of the order of the appearance of the stomata on the photosynthesizing organ, two main types of development can be distinguished: i) that in which the stomata appear gradually in a basipetal sequence, i.e. from the tip of the organ to its base, as is the case in leaves with parallel venation and in the internodes of the articulated species of the Chenopodiaceae; ii) that in which there is no regularity in the appearance of the stomata in the various regions of the growing organ, as is the case in leaves with reticulate venation. Stomata are distributed at more or less equal distances, the extent of which is specific to the plant species and leaf side. Several theories were suggested to explain the presence of the stomata-free regions surrounding each stoma: i) inhibition of additional stomata by already differentiated ones; ii) formation of the stoma together with the surrounding cells as part of the same developmental pattern; iii) induction of stomata pattern by the pattern of the underlying tissue.

Epidermal appendages

Massive structures such as warts secretory structures, and spines, which consist of epidermal as well as subepidermal tissues, are termed *emergences*. In some cases it is difficult to distinguish clearly between these two types of appendage without an ontogenetic study. For practical reasons, the appendages discussed below, either of epidermal or epidermal and subepidermal origin, will be termed trichomes.

The cells of the trichomes may develop secondary walls, which in some cases may become lignified. Some trichomes may lose their living protoplasts. The use of trichomes in taxonomy is well known. Some families can be easily identified by the presence of a particular type or types of hair. In other cases the hairs are important in the classification of genera and species and in analysing interspecific hybrids. Simple unicellular or multicellular uniseriate, non-flattened hairs, as are common, for example, in the Lauraceae, Moraceae, *Triticum, Hordeum, Pelargonium,* and *Gossypium.*

In *Gossypium* the fibres used in commerce constitute unicellular epidermal hairs, which may be up to 6 cm long and which are located on the seed coat. This group includes papillae and bladders, which are also known as *vesicular hairs.* Squamiform hairs which are conspicuously flattened and multicellular. These may be sessile and are then termed scales, or stalked, and known as peltate hairs, e.g. in *Olea* or dendroid or dendritic, e.g. in Cruciferae.

Branched, multicellular hairs which may be stellate, e.g. in *Styrax*, candelabrum-like, e.g. in *Platanus* and *Verbascum.* Shaggy hairs, multiseriate coarse trichomes, which consist, at least at the base, of two or more contiguous rows of cells. Such hairs may be seen on the petiole base of *Portulaca oleracea*, in *Schizanthus*, and certain species of the Compositae. In some species the hairs may show movements. This may be brought about in two ways: either by hygroscopic mechanisms, i.e. by the differential swelling and shrinking of the cell walls or by the action of living cells which may comprise the hair itself or which may be present only at the base of the hair or close to it. Glandular trichomes are involved in the secretion of various substances, e.g. salt solution, sugar solutions, terpenes, and gums. The secreting trichomes are often called glands.

The root hairs are usually viable for only a short period, generally only a few days. With the death of the root hairs and if the cells are not sloughed, the walls of the epidermal cells become suberized and lignified. In some plants root hairs have been found that remain permanently on the plant, The walls of such root hairs become thick and then apparently lose their ability to take up water from the soil. Both ontogenetically and functionally the epidermis may be considered as a separate tissue. The epidermis develops from the protoderm by continuous anticlinal cell divisions. As it is a compact tissue, devoid of intercellular spaces and covered with a cuticle, the epidermis provides protection to all those plant organs that consist entirely, or almost so, of primary tissues.

The specialized cells of the epidermis are of great interest as they usually are of characteristic structure, ontogenetic development, and function. They are also of great value in the study of taxonomic and evolutionary problems. Specialized cells, such as the trichoblasts and guard cells, are the products of unequal cell divisions, and they arise from the smaller of the two cells thus formed. Bunning has already drawn attention to this phenomenon which is characteristic not only for the epidermis but also for idioblasts, such as raphide-containing cells and idioblastic sclereids, which develop in inner tissues.

Winning states that the small embryonic cells, which are capable of varied development, are at first qualitatively equal and that their consequent development is, therefore, controlled by factors which as yet have not been analysed. Till now no satisfactory explanation has been given of the phylogeny and functional significance of certain of the specialized cells, such as the bulliform, silica- and cork-cells of grass leaves. Severai types of arrangement of the subsidiary cells around the guard cells are recognized in both monocotyledons and dicotyledons.

In relation to the dicotyledons, no conclusions have been put forward as to the evolutionary trends among these types. For the monocotyledons, Stebbins and Khush suggest that the type with four or more subsidiary cells is the most primitive, and that those types with few subsidiary cells or those lacking them entirely have been derived, independently, from this primitive type by reduction. The Stebbins and Khush approach was criticized by Paliwal. The argument for numerous subsidiary cells as the primitive type in monocotyledons was also not supported by the study of the development of the stomatal complex in representatives of 48 families carried out by Tomlinson and by the work on the subsidiary cells of Williams. It should also be mentioned that this evolutionary trend does not correlate with that of other anatomical characteristics such as the evolution of the tracheary elements, for instance. It is seen, therefore, that additional research should be made in this field. As regards the opening and closing of the stomatal aperture, the structure of only a few types of guard cell has been thoroughly investigated. There are indications that the variations in guard-cell structure result in different opening and closing mechanisms.

Chapter 4

Anatomy of Stem

Stem is the vegetative body of plant. The bud of the shoot usually consists of an axis, the *epicotyl,* containing a few internodes, which have not elongated, and some leaf primordia. With the germination of the seed the embryo enlarges and starts to grow, the apical meristem of the young shoot adds further leaf primordia, and the internodes between the lower primordia, which in the meantime have become distant from the apex, elongate. Buds develop in the axils of the developing leaves giving rise to a branched shoot.

In mature plants the development of leaf primordia at the shoot apex and the elongation of the nodes below it are the same as that in the growing embryo of the germinating seed. The order of appearance and the arrangement of leaves on the stem are more or less characteristic of each species. That part of the stem from which a leaf or leaves develop is called the *node* and that portion of the stem between two such nodes, the *internode.* The length of the internodes varies in the different species.

In certain plants, such as *Chichorium, Thrincia,* and others, where the leaves are arranged in a basal rosette, the internodes hardly elongate at all, but in most spermatophytes the internodes elongate to differing extents. At each node one, two, or more leaves may be found. The arrangement of the leaves on the stem is termed *phyllotaxy* or *phyllotaxis.*

When there are more than two leaves at one node the arrangement is termed *whorled.* When there are two leaves at each node the leaves are said to be *opposite;* in this type of arrangement the leaves of the successive nodes may be at right angles to each other and then the arrangement is termed *decussate;* or the leaves may form two parallel ranks along the stem, i.e. *distichous.* The straight vertical line along which the leaves occur is termed *orthostichy.* When there is a single leaf at each node and the leaves are arranged spirally on the stem, the phyllotaxy is said to be *alternate.* The helix along which the leaves occur is termed *parastichy,* The space, on the circumference of the stem, between two successive leaves, whether they arise at a single node or whether they are arranged spirally on the stem, is constant.

The position of the leaf primordia on the stem apex is determined before it is possible to distinguish any feature which indicates that such development has begun. Therefore it appears that the factors determining the position of the primordia on the stem apex are internal and, in general, they are identical with those factors that control the distribution of the growth potential in the apical meristem. The determination of leaf arrangement has been ascribed either to interactions within the apex or to the influence of the mature tissues below the apex, via the pro cambium. The evidence in favour of local determination within the meristematic tissues comes from studies of regeneration after wounds.

Three main theories have been the basis of much research concerning the possible nature of these local interactions within the apex:

i) The theory of the *first available space,* according to which a leaf primordium arises in the first space which attains a minimum width and a minimum distance below the tip of the shoot apex;

ii) The theory of *leaf fields* or *primordial fields,* according to which the existing primordia together with portions of the apical meristem form physiological units, new primordia being formed at specific points with reference to these fields;

iii) The theory of multiple foliar helices, according to which special mitotic properties are transmitted acropetally along foliar parastichies which terminate in leaf-generating centres that function in the "anneu initial".

The primary body of stem develops from the protoderm, procambium, and ground meristem. The arrangement and structure of the primary tissues are described below.

Stem cortex

The cylindrical region between the epidermis and the vascular cylinder is known as stem cortex. It may comprise various cell types. In the simplest case, the cortex consists entirely of thin-walled parenchyma tissue. In many stems, as for instance *Pelargonium, Retama,* and *Salicornia,* this parenchyma may have a photosynthetic function in addition to that of the temporary storage of starch and other metabolites. In other cases the outer region of the cortex, which borders on the epidermis, may include collenchyma or fibres, and the inner region, parenchyma. The collenchyma or fibres may form a continuous cylinder or they may be present in the form of separated strips.

Endodermis

Morphologically, a well-developed endodermis is a compact layer of living cells which form a hollow cylinder. The walls of the endodermal cells are of characteristic and specialized structure. In the typical cells, bands or strips which contain lignin and suberin, apart from cellulose, are present on the radial and transverse walls; these are called *Casparian strips* or *Casparian bands*.

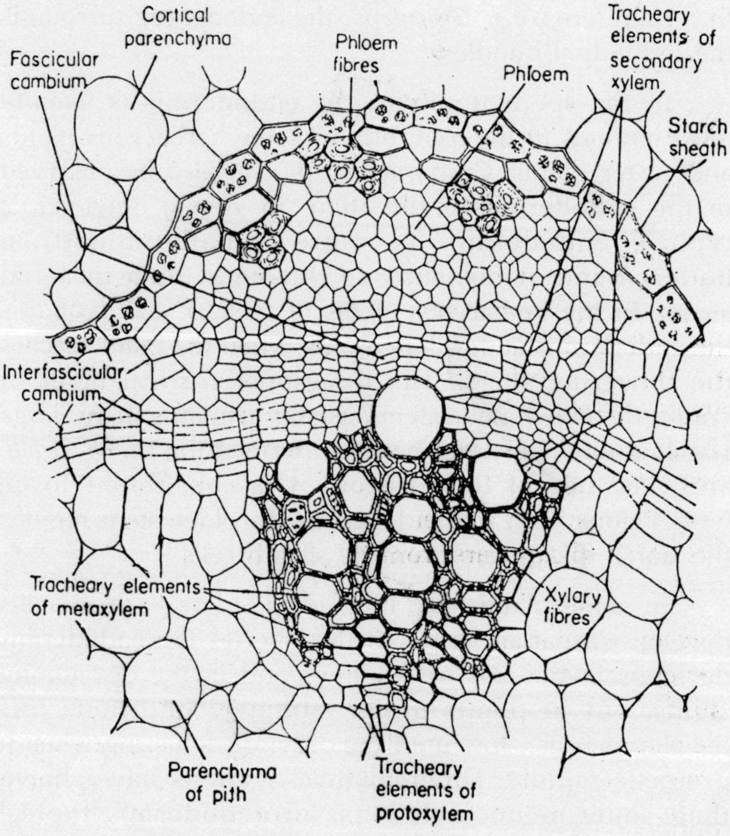

Fig. 1. Cross-section of a hypocotyl of Ricinus

In the course of maturation the endodermal cells may undergo changes that involve the addition of suberin lamellae on the entire inner surface of the cell walls. This may be followed by the addition of a secondary layer of cellulose, which may sometimes contain lignin, on the inside of suberin lamellae. The endodermis is most conspicuous in the stems of the lower vascular plants. Here the cell walls have Casparian strips and a suberin lamella, and the endodermis surrounds the vascular tissue. It is sometimes also found between the vascular cylinder and the pith *(Marsilea, Ophioglossum,* and others). In certain ferns, e.g. *Dryopteris,* the endodermis surrounds the individual bundles.

In the spermatophytes the endodermis is usually most obvious in the root, but, in some herbaceous plants and water plants, Casparian strips can also be observed in the endodermis of the stem. In stems, however, a typical endodermis is more usually found in underground stems, such as rhizomes, than in aerial stems. In the herbaceous stems of *Senecio* and *Leonurus* the endodermis is developed only when the plant reaches the flowering stage. The innermost cortical layer of young dicotyledonous stems usually contains many large starch grains. This layer has been termed the *starch sheath,* and, because of its position, it is considered to be homologous with the endodermis. In older stem regions the starch disappears from the sheath cells.

In some plants, e.g. in *Senecio vulgaris,* these cells develop Casparian strips. In this species the endodermis develops in the axis of seedlings and in the developing stems, but in plants under summer conditions this development is, slow until the onset of flowering when it proceeds rapidly. Different investigations have shown that, some monocotyledons, an endodermis typical secondary cell walls may be induced to a greater extent under the influence of external factors, such as the lack of nitrogenous salts and a high degree of aeration of the soil, among others.

In some dicotyledonous stems Casparian strips may develop under conditions that cause etiolation. In the stems of the lower vascular plants and in the roots of all vascular plants, the pericycle can readily be located between the endodermis and the vascular tissues. When dealing with stem anatomy of angiosperms, Blyth suggested to avoid the term pericycle, because in several species she has studied, the so-called pericyclic fibres have been found by her to be primary phloem fibres.

Primary vascular system

In the gymnosperms and most of the dicotyledons the vascular system consists of a continuous or a split cylinder which, encloses the *pith,* i.e. the central portion of the stem. In this cylinder two types of vascular tissue can be distinguished—the phloem which is usually external, and the xylem which is usually internal. In the case of the split cylinder each strand is termed a *vascular bundle.* A vascular bundle in which the phloem is only external to the xylem is said to be a *collateral bundle.*

In some dicotyledonous families, e.g. the Solanaceae, Cucurbitaceae, Asclepiadaceae, Apocynaceae, Convolvulaceae, and Compositae, internal phloem is also present. The internal or intraxylary phloem may be present as separate strands on the border of the pith, as in *Lycopersicon,* or it may be in close contact with the inner side of the xylem as in the stems of the Cucurbitaceae and Myrtaceae.

The internal phloem may differentiate both from procambium and from partly differentiated parenchyma cells. A vascular bundle in which the intraxylary phloem is in close contact with the xylem is termed a *bicollateral bundle.* Anastomoses between external and internal phloem of small bicolateral leaf veins have been reported for *Ecbalium elaterium;* anastomoses between the phloem of neighbouring vascular stem bundles of many species of the Cucurbitaceae have also been reported.

Careful observation of cleared stem portions stained for callose has revealed that phloem anastomoses between vascular bundles may be common to other families as well. Transport of labelled compounds in intact tobacco plants showed that translocation in the internal phloem is of lesser importance than that of the outer phloem. Botha and Evert arrived at a similar conclusion regarding leaf veins of *Cucurbita maxima.* These authors found that most of the stylets of aphids which were feeding on the leaves terminated in the abaxial phloem.

In stems of some Asclepiadaceae, however, most of the aphid stylets were observed to end within the internal primary phloem. In some monocotyledons, as, for example, *Convallaria majalis, Acorus,* some genera of the Xanthorrhoeaceae, as well as the secondary bundles of *Aloe arborescens, Dracaena, Cordyline,* and others, the xylem surrounds the phloem. Such bundles are termed *amphivasal bundles.* Bundles common in the Pteridophyta, in which the phloem surrounds the xylem, are termed *amphicribral bundles.*

There are also bundles in which the xylem is seen to be V- or U-shaped in cross-section. In the former, as can be seen in the leaf of *Xanthorrhoea,* for example, the phloem groups are situated at the free ends of the xylem arms. In the case of U-shaped xylem the phloem is surrounded on three sides by xylem; such bundles occur in the stem of *Asparagus aphyllus* and *Kingia australis.* In most monocotyledons and in a few dicotyledons no distinct vascular cylinder exists. The primary vascular system consists of a large number of bundles which are scattered irregularly, and it is impossible to distinguish clearly the boundary between the cortex, the vascular cylinder and the pith. The vascular system plays an important role in the attempt to solve phylogenetic problems. It hits been thoroughly investigated, by comparative method, both in plants living today and in fossil forms. Van Tieghem and Douliot proposed the *stelar theory* to explain the structure of the plant axis. This theory greatly influenced later investigators who worked on the comparative anatomy of the Tracheophyta.

According to the above theory the gross anatomical structure of the root and stern is similar, i.e. in both of them the cortex Surrounds a central core which they termed the *stele.* The stele comprises the *pericycle,* i.e. the non- vascular tissue between the phloem and cortex, the *vascular tissues* and the *pith,* when present. Medullary and cortical bundles, present on the inside and outside of the stele respectively, are found in certain plants. These bundles are associated with stems of both anomalous and typical structure. Medullary bundles occur in numerous dicotyledonous families, e.g. in the Amaranthaceae, Chenopodiaceae, Orobanchaceae, Berberidaceae, and Cucurbitaceae. Cortical bundles are less common and are known to occur, for example, in the Melastomaceae, Proteaceae, Araliaceae, and Calycanthaceae.

Pattern of arrangement of the primary vascular tissues differs in the various plant groups and sometimes even within the same species. It has been found that the different patterns illustrate various stages in the evolution of the primary vascular system. The pattern in each plant is determined in the shoot and root apices from which the vascular tissues differentiate according to its genotype. The stele of the sporophyte of vascular plants may be divided into two basic type: i) *protostele* which consists of a solid central cylinder of xylem surrounded by phloem; and ii) *siphonostele* in which there is a cylinder of pith within the xylem. Both ontogenetically and phylogenetically the protostele is the more primitive, and it is thought that the siphonostele has developed phylogenetically from the protostele. Three types of protostele call be distinguished: a) *haplostele* which is the simplest type, in which the xylem appears more or less circular in cross-section, *Rhynia* and *Selaginella;* b) *actinostele* in which the xylem is stellate in cross-section, e.g. *Psilotum:* and c) *plectostele* in which the xylem is split into longitudinal plates of which some are joined and others separate, e.g. *Lycopodium.*

Two types of siphonostele are distinguished according to the positions of the phloem and xylem: a) *ectophloic siphonostele* in which phloem only surrounds the xylem externally: and b) *amphiphloic siphonostele* in which the phloem surrounds the xylem both externally and internally and where the endodermis appears both outside and inside the vascular tissue on the borders of the cortex and pith, respectively The siphonostele may consist of a continuous cylinder of vascular tissue or of a network of bundles. The latter type is the more advanced and the vascular tissues of this type appear in cross-section as a ring of separate bundles.

Those parenchymatous regions that occur in the stele, above the positions where the leaf traces diverge from the stele to the leaves, are termed *leaf gaps.* An amphiphloic siphonostele in which the successive leaf gaps are considerably distant, one from the other, is termed a *solenostele.* An amphiphloic siphonostele with overlapping gaps, i.e. that in which the lower part of one gap is parallel with the upper part of another gap, is termed a *dictyostele*. In this case the bundles are interconnected to form a cylindrical network, and each bundle is of concentric structure consisting of a central strand of xylem surrounded by phloem. Individually such bundles are termed *meristeles.* From the anatomical point of view these are amphicribral bundles. Bicollateral vascular bundles in which the xylem strands are accompanied externally and internally by phloem strands and which are found in advanced dicotyledonous families appear to be the result of a secondary specialization and not a relic of the primitive structure characteristic or the Filicinae.

The vascular system of eustelic plants composed of bundles that branch and anastomose is referred to as a closed system. A vascular system characterized by bundles which branch but still form independent, separate systems, is referred to as an open system. A compact arrangement of vascular bundles, which may form an open or closed system, seems to correlate with woody habit. A different interpretation of the above nomenclature exists and has been summarized by Sporne.

According to this interpretation the ectophloic siphonostele without leaf gaps, such as is found in the Pteridophyta, is considered as a protostele and is termed a *medullated protostele.* Sporne does not use the term siphonostele for the Pteridophyta. As in the dictyostele, the bundles of the eustele are usually interconnected. That type of stele in which the bundles are scattered, such as is characteristic of the monocotyledons, is called an *atactostele.* In the siphonostele not all the interruption in the vascular tissue are leaf gaps as described above. Some interruptions result from the secondary reduction of vascular tissue and the formation of interfascicular parenchyma. Such interruptions are termed *perforations*. When such perforations occur in a solenostele it may be confused with a dictyostele. The parenchymatous connections between the pith and cortex are termed *medullary*. Differing from the Pteropsida, whose steles form leaf gaps, the steles in microphyllous plants, i.e. the Psilopsida, Lycopsida, and Sphenopsida, are devoid of such parenchymatous regions. In microphyllous plants with a siphonostele the only gaps present are *branch gaps* which are those gaps associated with bundles that depart from the central cylinder to the lateral branches. This type of siphonostele has been termed by some authors *cladosiphonic,* and that found in the Pteropsida *phyllosiphonic.*

The siphonostele developed from the protostele by the transformation of the inner Initials of the tracheary elements into parenchyma initials. This view is supported by the fact that tracheary elements may be found in the centre of the stele scattered among the parenchyma cells. This feature is common in relatively primitive plants, both living and fossil. Such steles may be considered as intermediate stages between protosteles and siphonosteles.

Research on *Ophioglossum lusitanicum,* for instance, showed that the stele of the rhizome of the sporophyte, of this species, that developed from the gametophyte was protostelic at the base and siphonostelic in its upper portion. Furthermore, it was clearly seen in the transition zone, below the level of the first leaf gap, that the pith undoubtedly originated from the xylem. Here parenchyma cells were seen mingled with tracheids and the number of parenchyma cells was seen to increase in an upwards direction. The vascular systems discussed above are composed entirely of primary tissues, i.e. of protoxylem and metaxylem and protophloem and metaphloem. The order in which successive metaxylem elements mature may be centrifugal or centripetal.

When the protoxylem is on the inside and the differentiation of the metaxylem proceeds progressively towards the periphery, as in angiosperm stems, the xylem is described as *endarch.* When the protoxylem is on the outside and the metaxylem differentiates centripetally, as in angiosperm roots, the xylem is described as *exarch.* In a third arrangement termed *mesarch,* the differentiation of the metaxylem proceeds both centripetally and centrifugally to the protoxylem. The exarch and mesarch types of primary xylem are apparently the more primitive. The phloem always develops towards the xylem. In those cases where the phloem is situated only on the outside of the xylem the development of the phloem is centripetal, and in internal phloem the development is centrifugal.

In the gymnosperms the evolution of the stele proceeded, as suggested by sporne, as follows: solid protostele with mesarch protoxylem, split tracheary tissue mixed with parenchyma, with protoxylem only in the outermost strands; mesarch xylem strands Surrounding a pith, endarch xylem strands surrounding a pith. The direct evolution of the eustele of the gymnosperms from the protostele by gradual medullation and concurrent separation of the peripheral conducting tissue into sympodial bundles was suggested also by Namboodiri and Beck. The closed vascular system of conifers is regarded by these authors as derived from the open sympodial system characteristic of most gymnosperms, by fusion of pairs of leaf traces.

In dicotyledons, the primary vascular cylinder is interrupted at each node by the exit of one or more bundles that enter the leaves. The stelar bundles, which are the continuation of the bundles in the leaf bases, are called *leaf* traces. According to the number of leaf gaps per leaf the node is, termed *unilacunar, trilacunar* or multilacunar. The anatomy of the node is an important aspect of taxonomy and of the comparative morphology of the stem, leaf, and flower.

The trilacunar node is the primitive type in the angiosperms and that the unilacunar node developed, phylogenetically, from it by the loss of the two lateral gaps together with their respective traces, or by the approximation of the lateral traces to the median bundle to form a bundle composed of three traces, which is associated, therefore, with a single gap. Contrary to the reduction in the process of the formation of the unilacunar node.

The multilacunar node is formed by the addition of new gaps and traces. The assumption that the trilacunar node is the primitive one among the angiosperms has been refuted by later research on nodal anatomy in the gymnosperms and angiosperms. By these workers and others it has been shown that in many Pteridophyta, in the Cordaitales, Bennettitales, in *Ginkgo and* Ephedra, a single gap is found in that position where the leaf traces depart from the stele. Similarly, in many dicotyledons unilacunar nodes with two leaf traces have also been found.

Many of the dicotyledonous genera with unilacunar nodes with a double leaf trace belong to the primitive groups of the Ranales and the Chenopodiaceae. Bailey found in many dicotyledons that the vascular supply to the cotyledons consists of a double leaf trace which arises from a unilacunar node. Bailey is of the opinion that the leaves of angiosperms are able to undergo reversible changes in shape and vascularization. In the light of the above facts it can be assumed: i) that the unilacunar node of certain genera of the Ranales is primitive and has not changed during its evolution, from that of the lower Pteropsida; ii) in certain other dicotyledonous genera, e.g. genera of the Leguminosae, Anacardiaceae, and others, the Unilacunar node has apparently been derived, by reduction, from a trilacunar node; iii) there are positive indications in certain dicotyledonous groups, such as the Epacridaceae and Chloranthaceae, that in the course of evolution the tri- and multilacunar nodes have arisen from the unilacunar node.

Evolutionary development from a unilacunar node with two leaf traces to other types of unilacunar nodes with one, three, or more traces may occur in a single family as can be demonstrated in the Chenopodiaceae. In the Gnetaceae, *Ephedra* has unilacunar nodes with two leaf traces whereas in *Gnetum* the nodes are multilacunar.

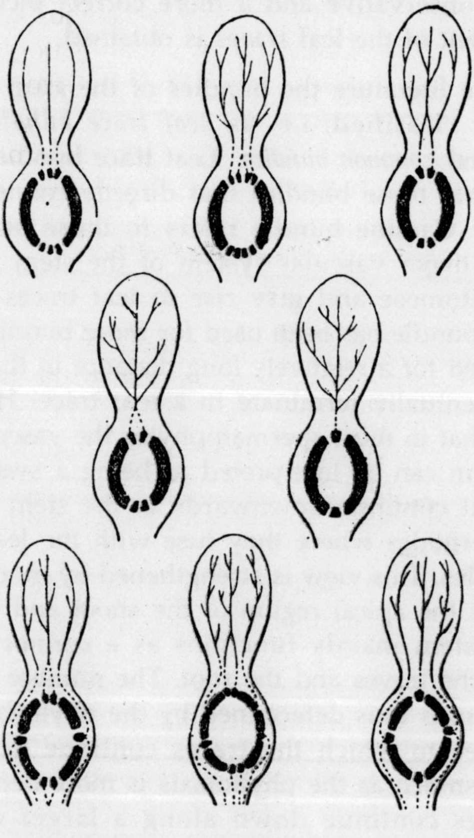

Fig. 2. Different types of vasculatisation of cotyldonary node in dicotyledons

As stated by Rodin and Paliwal the nodal anatomy of *Gnetum* represents the highest degree of specialization among the gymnosperms. In order further to clarify the nature of the leaf traces and the nodal type, it is necessary to study more accurately the nature of the traces and to follow their passage downwards in the stem. Often the picture obtained at the node is not a true reflection of the situation, but lower in the stem the stele is more conservative and a more correct picture of the arrangement of the leaf traces is obtained.

In the literature the bundles of the stem have been variously classified, i.e. as *leaf trace bundles, cauline bundles,* and *common bundles.* Leaf trace bundle was used to designate those bundles that directly connect the leaf and stele. Cauline bundle refers to those bundles that form the major vascular system of the stem and which may anastomose and give rise to leaf traces. The term common bundle has been used for those bundles that run unbranched for a relatively long distance in the stem and which eventually terminate in a leaf trace. However, it appears that in most spermatophytes the vascular system of the stem can be interpreted as being a system of leaf traces that continue downwards in the stem for one or more internodes where they fuse with the leaf traces of lower nodes. This view is strengthened by morphogenetic studies of the apical region of the shoot and by the fact that the stem mainly functions as a connecting organ between the leaves and the root. The number of bundles in the stem is thus determined by the phyllotaxis and by the degree to which the traces continue down in the stem. Inasmuch as the phyllotaxis is more dense and the leaf traces continue down along a larger number of internodes, the number of bundles, as seen in a cross-section of the stem, will be greater. Changes in number may sometimes occur along the stem of one plant.

Branches develop from axillary buds and have vascular connections with the main axis. These vascular connections are termed *branch traces.* At the node, the branch traces are situated very close to the leaf traces that enter the leaf in whose axil the branch develops. The branch gap is situated above the leaf gap and together these two gaps appear as one. In the dicotyledons and gymnosperms the vascular supply to the axillary branches usually consists of two traces. However, in some plants only one trace is present and in others more than two. In the branches themselves, the stele is similar to that of the main axis.

The vascular system of monocotyledons usually consists of bundles that are scattered throughout the ground tissue of the stem. In cross-sections of such stems it can be seen that the bundles do not form a ring such as is found in cross-sections of most dicotyledons. Among dicotyledons, however, an arrangement of more or less scattered bundles does occur in a small number of plants, e.g. in the Nymphaeaceae, many plants of the Ranunculaceae and the herbaceous genera of the Berberidaceae.

The course of vascular bundles in the stem of palms was investigated by Zimmermann and Tomlinson by a cinematographic method. According to these studies essentially all bundles maintain their individuality and proceed indefinitely up the stem" from the base of the stem to the apex. Each of these vertical bundles—minor, intermediate, and major—give off a leaf trace at intervals. The minor bundles give off leaf traces more frequently than the major ones. Each vertical bundle slowly reaches a more central position, when followed from the base to the apex.

At intervals the bundle "turns" sharply towards the periphery whilst it gives off the leaf trace, satellite bundles which go into the inflorescence and bridge bundles which make connections with neighbouring bundles. Major bundles extend more into the centre of the stem than the minor one. Because only relatively few bundles extend to the centre of the stem and because all bundles in parts of their course are found in the peripheral region, this region is crowded with bundles. In the central, uncrowded part of the stem, all bundles when traced upwards rotate uniformly and describe a helix in the direction of the phyllotactic spiral.

The vertical bundles described by Zimmermann and Tomlinson could be regarded as sympodia consisting of leaf traces. The primary vascular system of arborescent Liliflorae is in principle similar to that of the palms. In other monocotyledons, e.g. in the Dioscoraceae, the vasculature of the stem is very complicated and differs markedly from that of the palms.

The complicated arrangement of vascular bundles in the monocotyledonous stem, and especially in the Gramineae, is connected with the position of the various bundles of each leaf in the stem. According to Kumazawa, three types of leaf trace bundles are present in the stem of Zea. The first type is represented by those bundles derived from mid-ribs and the larger lateral veins. These penetrate deep into the interior of the stem and then pass to the periphery in lower nodes. The second type is represented by leaf trace bundles derived from smaller lateral veins which, immediately on entering the stem, occupy the position of the outermost peripheral bundles with which they fuse sooner or later. The third type is represented by bundles that are very thin and which soon fade out within the cortex close to the level of the node.

In the Gramineae there are special horizontal bundles at the nodes which connect the leaf bundles together. In the stem of *Triticum* the internodes are hollow and the nodes solid. The thin leaf bundles are continued in the outer circle of the stem, and the thick bundles constitute the inner circle of bundles. In the nodes of *Triticum* O'Brien and Zee distinguished three types of vascular bundle: i) nodal plate bundles which contain phloem transfer cells; ii) large bundles elliptical in cross-section, containing xylem transfer cells; and iii) diffuse bundles with irregular outline, which are apparently a result of fusion of collateral bundles of the internode, and which alternate with the elliptical bundles.

With the progressive differentiation of the promeristematic cells the three meristems, the protoderm, ground meristem and procambium, are formed. The procambium may be in the form of a solid or hollow cylinder or of strips. The developing procambial strips, as seen in cross-sections, are organized in groups of cells which divide in all directions, the commonest direction being periclinal. The last gives the cambial strand a concentric appearance. The procambium stains more intensely than the neighbouring tissues as a result of the delayed appearance of the vacuoles in it. The development of the procambium is continuous in an acropetal direction, i.e. from below towards the apex.

The procambium differentiates below the position where a leaf primordium is formed as a small buttress on the surface of the apex. The differentiation of procambial cells into cells of the vascular tissue takes place in different plants and organs at different stages of procambial development. For instance, in aerial stems of the spermatophytes the procambial cells undergo differentiation into tracheary elements even at the level where cell division still takes place in the procambium, while in the roots of most spermatophytes and in the stems of most pteridophytes the first differentiation of the procambium takes place only where almost all of the procambium cells have ceased to divide.

Earliest sieve elements usually undergo acropetal differentiation, i.e. from the phloem of the traces of more mature leaves, to the leaf primordia. The development of the phloem commences before the appearance of the xylem elements. In spermatophytes the primary xylem begins to differentiate from the procambium near the base of the developing leaf, from where its further differentiation continues in two directions—acropetally into the leaf, and basipetally into the stem. In the stem the newly formed xylem becomes continuous with the xylem of earlier formed strands. The above processes of differentiation lime been proved only in relation to the development of the protophloem and protoxylem.

As yet the pattern of differentiation of the metaphloem and metaxylem along the shoot has not been investigated thoroughly in the spermatophytes. In the procambium it is possible to distinguish between that part that will differentiate into xylem and that part that will differentiate into phloem by the more intense staining of the latter as well as by the differences in the plane of cell division. In plants that exhibit secondary thickening that part of the procambium between the primary xylem and phloem remains meristematic and forms part of the vascular cambium.

Experimental work on the control of vascular tissue differentiation has yielded some understanding of the mechanisms determining the pattern of vascular strands. It has been shown, both in plants and in tissue culture, that young leaves cause the differentiation of xylem and this effect can be replaced by auxin. The differentiation of vascular strands appears to occur along the path of auxin transport, from the shoot to the root.

Differentiation into defined strands, rather than into a diffuse system, can be explained by polarization of auxin transport occurring at an early stage in vascular differentiation. The cells which first start to differentiate, therefore, become the preferred channels for auxin transport, and this in turn causes their further differentiation and diverts the auxin from cells which are not within the differentiating strand. Leaf removal and hormone treatment experiments showed that contacts between strands are readily formed only between a new strand and an established strand which does not connect to a young leaf and, therefore, is not loaded with auxin.

Pith

A pith is a more or less cylindrical body of tissue in the centre of the axis, enclosed by the vascular tissues. It has long been known under the name of *medulla* which was later abandoned, although the extensions of the pith between the vascular bundles have retained the name *medullary rays.* Pith consists of a rather uniform tissue, mainly parenchymatous, in which the cells are arranged usually loosely. Often thick-walled, lignified parenchyma cells and sclereids are also present. Fibres occur only rarely, and if so, in the peripheral region where they are associated with the primary vascular tissue. In some species secretory structures occur in the pith.

In the ontogeny of the stem, pith cells in the internodes of many species mature very early and stop growing, whereas the surrounding tissues are still meristematic and continue to enlarge longitudinally and in circumference. Thus the pith may be torn apart and a hollow pith is formed, with the broken cell walls lining the cavity. This is common in herbaceous plants.

Stem adaptations

The development of storage organs from stems results from the activity of primary or secondary meristems. In kohlrabi, for instance, this occurs between the third and fifth nodes. The meristematic activity resulting in bulbing of the stem initiates in the central pith, after a normal cambial zone has been established. The restricted penetration of light into water creates conditions similar to those in which densely shaded land plants grow. The submerged leaves and stems have an increased distribution of chloroplasts and a reduced cuticle. There is a high surface area/volume ratio in the leaves, which are often dissected or fenestrated.

The stems contain an extensive system of intercellular spaces through which diffusion of dissolved gases may freely occur. Absorption of gases is also facilitated by the thin walls of the epidermis and inner tissues. Cuticle is usually absent from submerged organs, but a very thin cuticle was reported to occur on the epidermis of stems and leaves of several species, e.g. *Elodea canadensis, Groenlandia densa, Hottonia palustris,* and *Myriophyllum spicatum.* Chloroplasts generally occur throughout the mesophyll of leaves and outer cortex of many stems. In submerged leaves and stems of water plants chloroplasts occur in the epidermal cells, a feature common also to dicotyledons of deeply shaded habitats. In *Ceratophyllum, Myriophyllum, Potamogeton,* and some marine angiosperms, e.g. *Zostera,* the epidermis is richer In chloroplasts than the inner tissues. In these species the cortex and mesophyll function mainly as storage tissues for starch and lipids. In the Podostemaceae chloroplasts also occur in the epidermis of the flattened root thalli. In most submerged hydrophytes the epidermis lacks stomata.

In the stem cortex and in the ground tissue of the petioles and mesophyll except for the very reduced leaves, schizogenous air passages of lacunae exist. The lacunae occur in the middle cortex of the stem. The outer cortex consists of a compact parenchyma or collenchyma. The inner cortex which surrounds the vascular cylinder also consists of a dense parenchyma.

The lacunae may be arranged in one ring, e.g. in. *Ceratophyllum* and *Myriophyllum* or in several rings, e.g. in *Hippuris*. In the petiole of *Nuphar* the lacunae are arranged in a reticulate pattern. In some genera specialized cells occur in the tissue surrounding the lacunae. In *Myriophyllum,* cells with clustered crystals of calcium oxalate are found. In *Brasenia* mucilage glands project into the lacunae. In *Nuphar, Nymphaea,* and *Victoria* sclerenchymatous idioblasts of bizarre shape occur. The lacunae are transversed at intervals by plates, or diaphragms, which strengthen the organs and may also eliminate the danger of waterlogging through wounds.

Diaphragms in stems are most commonly found at the nodes. The diaphragms vary in structure and development. In *Nuphar* the lacunae are at first continuous, then groups of cells surrounding the lacunae at certain places swell, branch in the cavity, and divide to form a spongy mass which forms a diaphragm. In *Pontederia, Sagittaria, Potamogeton* and some other genera, the diaphragms are one or sometimes two to three cells thick. Between the cells characteristic intercellular spaces develop especially at the cell angles. In non-submerged aquatic plants, e.g. in *Cyperus* and *Scirpus*, the spaces between the diaphragms are filled with a stellate parenchyma. The most specialized tissue found in the stem and respiratory roots of many water plants is the *aerenchyma*.

Aerenchyma is, strictly speaking, a phellem tissue derived from a typical phellogen of either cortical or epidermal origin. During the development of this phellem, single cells from each cell layer elongate in a radial direction at constant intervals, while the cells between them remain short. This lengthening of the cells pushes the previously formed layer of cells outwards so that numerous elongated air chambers, parallel to the axis of the plant, are formed. The elongated cells form the radial walls of the air chambers, while the non-elongated cells form the tangential walls.

However, any tissue that contains large intercellular spaces is termed aerenchyma. Aerenchyma, which serves as an air transport system in waterlogged plants, may develop in crop plants growing in wet soil. The intercellular spaces of such plants develop mostly by disintegration of cells. It was suggested that in some plants, when waterlogged the deficiency of oxygen triggers anaerobic stimulation of ethylene production, which causes an increase in cellulase activity. This activity brings about disintegration of cells and thus the development of aerenchyma. As only part of the cells in the tissue disintegrate, Kawase suggests that there is competition for water between neighbouring cells after the cell walls have been softened by the action of cellulase. Stronger cells survive and become larger whereas weaker ones loose water, become plasmolysed, and die.

The root system of hydrophytes is usually, much reduced, it principally provides an anchorage in the soil since the uptake of water and salts is carried out by the stems and leaves. For the same reason the vascular system is also much reduced. The reduction is especially obvious in the xylem tissue. The roots of *Thalassia,* for instance, have no tracheary elements except close to the region of their insertion. An outstanding character of the submerged organs of water plants is the reduction in number and degree of lignification of the tracheary elements. The tracheary elements usually have annular or spiral wall thickenings and in many species are reduced to tracheids.

Chapter 5

Leaf Anatomy

Morphologically and anatomically the leaf is the most variable plant organ. The collective term for all the types of leaves appearing on plants is the *phyllome.* There are different types of phyllomes such as foliage leaves, cataphylls, hypsophylls, cotyledons, and others. The foliage leaves are the principal photosynthetic organs. The cataphylls are the scales that appear on buds and on underground stems and their function is protection or the storage of reserve materials. The first, lowermost leaves of a side branch are termed *prophylls*; in monocotyledons only one prophyll is usually present and in dicotyledons, two. The hypsophylls are the various types of bracts that accompany the flowers and their function is, apparently, protection. Sometimes the hypsophylls are coloured and then their function is similar to that of petals.

The foliage leaves in certain angiosperm genera are sessile and consist almost entirely of lamina, but in most genera several distinct parts can be distinguished, i.e. the *leaf base, petiole,* and *lamina.* The shape, structure, and relative size of these parts differ and are used to classify the foliage leaves into different types. In many dicotyledonous genera two appendages, i.e. the *stipules,* develop at the leaf base. The stipules may be attached to the leaf base or they may be free appendages. Even when they are free appendages they develop as outgrowths of the leaf primordia. The vascular supply of the stipules is derived from the leaf traces. In some plants the stipules are green and resemble leaflets and have a photosynthetic function, but their main function is the protection of young developing leaves. Most plants with trilacunar nodes have stipules and those with multilacunar nodes have sheath-like leaf bases.

Leaves are divided into *simple* and *compound* depending on whether the stalk bears one or more leaflets. In the case of compound leaves the common stalk is termed the *rachis.* If the leaflets arise from the sides of the rachis, the leaf is said to be *pinnate,* and if all from one point, as rays, *palmate.* The margin of the leaf, and similarly the leaflets, may be entire or variously notched.

Although most leaves are dorsiventral appendages they may be flattened laterally as in *Iris,* almost cylindrical as in *Sansevieria cylindrica,* or may have bizarre tubular forms as in some insectivorous plants. In some plants, such as many species of *Acacia* of Australian origin, the lamina of the leaves is reduced, and from the remaining parts a flat, leaf-like photosynthetic organ is developed; this type of organ is termed a *phyllode.* Although pinnae are actually initiated during ontogeny they soon become arrested in their growth. In certain plants the stems are photosynthetic and have become flat, such organs are termed platyclades. When the platyclade appears very leaf-like, as in *Ruscus,* it is called a *phylloclade.*

Epidermis

The epidermis of leaves of different plants varies in the number of layers, its shape, structure, arrangement of the stomata, appearance and arrangement of trichomes, and occurrence of specialized cells. Because of the usually flat structure of the leaf, a distinction is made between the epidermal tissues of the two surfaces of the leaf; that surface of the leaf that is closer to the internode above it and which usually faces upwards is referred to as the *adaxial surface*, and the other surface as the *abaxial surface*.

Xerophytic plants develop structural characteristics that are adaptations to the habitats. Xeromorphism, however, is not confined to xerophytes, and not all xerophytes exhibit xeromorphic characters. One of the most obvious features of xeromorphic leaves is the small ratio of the external leaf surface to its volume. According to numerous workers the reduction of the external surface is accompanied by certain changes in the internal structure of the leaf as, for instance, the reduction in cell size, the increase in the thickness of the cell walls, the greater density both of the vascular system and of the stomata, and the increased development of the palisade tissue at the expense of the spongy tissue.

Xeromorphic leaves are often covered with trichomes. In xerophytes with succulent leaves water-storing tissue is developed. The lack of nitrogenous compounds and/or water in the soil often results in the appearance of xeromorphic characters such as the formation of thick walls and cuticle and the additional development of sclerenchyma.

A clear relationship can be established between the salinity of the soil and the appearance of succulent features in the plants growing in it. Intense illumination and the retardation of the water flow due to water deficiency apparently results in the increased development of palisade tissue. Some of the above changes, such as the increase in the number of stomata, allow a higher rate of gas exchange under conditions of favourable water supply.

Furthermore, the increased development of palisade tissue probably results in an increase of photosynthetic activity. The reduction in the size of the leaf is thought to be a feature correlated with the reduction of the rate of transpiration and in many places it is seen that plants with small leaves are commoner in dry habitats. In some cases the reduction of leaf size is connected with an increase in the total number of leaves on the plant. Thus, for example, the total external surface area of the entire foliage of certain coniferous trees is usually greater than that of many dicotyledons.

Trichomes are very common on xerophytes. If a species exists in both xeromorphic and mesomorphic forms, the former has a denser indumentum. In many xeromorphic plants, such as, for instance, *Nerium* and *Xanthorrhoea*, the stomata are sunken in depressions or grooves which are covered by trichomes. Environmental factors can influence the development of the cuticle and wax. In *Prosopis velutina*, Bleckmann *et al.* found that the cuticle of the leaflets of plants grown outdoors was more than ten times as thick as that of plants grown indoors.

Insolation of photosynthetizing plant organs in deserts and semi-deserts is an important factor in reducing both the absorptance of photosynthetically active radiation and the temperature of the exposed organs. Hairs or scales, epicuticular wax and resin may play a role in insolation and light reflectance. According to Shields, living trichomes, which themselves lose water, do not protect the plant from excessive transpiration as do dead trichomes which form protective layers.

As trichomes are more numerous on plants growing in dry habitats, and as they do not prevent excessive evaporation, shields suggests that it is possible that the trichomes are symptoms of water loss rather than being structures that function to reduce evaporation. In some cases pubescence might play a role in deterring predators. Mucilage wall layers of epidermal cells may be involved in water economy of leaves of xerophytes.

Volkens suggested that in some desert plants the stomata on the photosynthesizing plant organs become permanently closed during the summer season. This closure is caused by the additional thickening and cutinization of the guard-cell walls or by the blocking of the sunken stomata from the exterior by resinous masses (e.g. Pityranthus) or by wax layers (e.g. Capparis spinosa). In Anabasis articulata and other related species prominent wall thickenings have been seen to develop in the guard cells in the hot summer month.

The permanent closure of the stomata of desert plants during the dry season should be investigated experimentally, as this feature explains how those green parts of desert plants that do not dry out manage to retain their water content. In *Rumex acetosella* drops of resin or oil form in the epidermis and in the cells around the veins under drought conditions; this feature apparently hinders the passage of water. It is possible that this is also the function of the tannins and resins found in species of *Quercus* and *Pistacia* of the Mediterranean maquis.

Atay and Heilbronn suggested that vapour of ethereal oils may lower the evaporation and transpiration rates. Water in leaves is conducted not only by the veins and bundle-sheath extensions, but also by the mesophyll cells and epidermis. Water transport towards the epidermis is much higher through the palisade tissue than through the spongy parenchyma. In centric xeromorphic leaves the palisade cells radiate around the central vascular bundles and therefore, under favourable conditions of water supply, the transport of water from the bundles to the epidermis is enhanced.

The presence of intercellular spaces, especially between the palisade cells, however, limits water transport in the plane parallel to the leaf surface. The volume of intercellular spaces is smaller in xeromorphic leaves than in mesomorphic ones. However, the ratio between the internal free surface area of the leaf and its external surface is small in shade leaves and large in xeromorphic leaves. Similar results have been obtained by us from plants of various ecological types; e.g. in *Styrax officinalis*, which is a mesophyte, the ratio is 8.91, while in *Olea europaea* and *Quercus calliprinos*, which are xerophytes, it is 17.95 and 18.52 respectively.

The increase in the free internal surface area is due to the increased development of the palisade tissue. The latter is probably one of the reasons why, besides the increase in photosynthetic activity the rate of transpiration of xerophytes is under conditions of favourable water supply. In some xerophytes, and generally in halophytes, well-developed water-storing tissues occur in the leaves. Water-storage tissue in the leaf consists of large cells with large vacuoles containing a dilute and/or mucilaginous cell sap. These cells have a thin layer of cytoplasm lining the cell walls in which scattered chloroplasts may be found.

The osmotic pressure in the photosynthesizing cells is higher than in the non-photosynthesizing ones and when water is lacking they obtain their water from the water-storing tissue. As a result of this, the thin- walled water-storing cells shrink, but under favourable conditions of water supply they rapidly return to their former state. Hypertrophy of parenchyma cells of some coastal plants as compared with those of the same species grown in inland habitats is a well-known phenomenon. In the reduced leaves of *Salicornia*, broad and short tracheid-like cells are found among the palisade cells. The function of these cells has been variously interpreted by different workers. Duval-Jouve stated that they are filled with air.

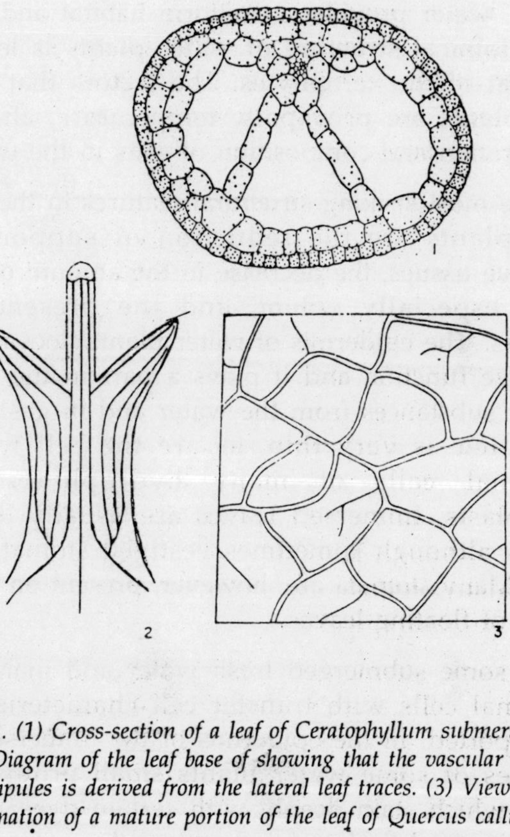

Fig. 1. (1) Cross-section of a leaf of Ceratophyllum submersum. (2) Diagram of the leaf base of showing that the vascular supply of stipules is derived from the lateral leaf traces. (3) View of the venation of a mature portion of the leaf of Quercus calliprinos.

According to Holtermann these cells transport water to the peripheral layers. Other investigators believe that these tracheid-like cells (tracheoids) have a water- storing function. *Tracheoid idioblasts* may be scattered throughout the mesophyll as, for instance, in *Pogonophora schomburgkiana* of the Euphorbiaceae. Involution of the leaves, which is especially typical of grasses, is a character of xerophytes. This feature is brought about by the action of bulliform cells and/or other epidermal and mesophyll elements which may be parenchymatous or sclerenchymatous.

In contrast with the different types of xerophytic habitats, water provides a uniform habitat and therefore the anatomical structure of water plants is less varied than that of the xerophytes. The factors that influence water plants are principally temperature, air, and the concentration and composition of salts in the water.

The most striking structural features in the leaves of water plants are the reduction of supporting and protective tissues, the decrease in the amount of vascular tissue, especially xylem, and the presence of air chambers. The epidermis of water plants does not have a protective function and it plays a part in the uptake of nutrient substances from the water and in gas exchange. The cuticle is very thin, as are the cell walls. The epidermal cells of many hydrophytes contain chloroplasts. Immersed leaves are usually devoid, of stomata although sometimes vestigial stomata may be found. Many stomata are, however, present on the upper surface of floating leaves.

In some submerged fresh water and marine plants epidermal cells with transfer cell characteristics have been reported. In the epidermis of the immersed surface of leaves of some water plants small groups of cells appear which stain deeply with certain dyes. These cell groups, termed *hydropotes,* are thought to be structures which facilitate water and salt transport into and out of the plants. In many species of water plants, e.g. *Ceratophyllum, Myriophyllum,* and *Utricularia,* the leaves are divided into narrow cylindrical lobes which greatly increase the area of contact with the water. In *Ranunculus aquatilis* the floating leaves are entire, while the immersed leaves are divided into narrow lobes.

In the leaves and stems of hydrophytes *air chambers,* filled with gases, are found. These chambers are intercellular spaces which are usually regular in shape and which pass through the entire leaf. In many plants the air chambers may penetrate deep into the tissues of the stem. This type of structure can be seen in the leaves of Potamogeton and *Eichhornia*. The air chambers are usually separated from one another by thin partitions of one or two layers of chloroplast-containing cells. Diaphragms—cross—partitions—occur in elongated air cavities. Such cavities are often termed air passages or lacunae. The diaphragms consist of a single layer of cells with small intercellular spaces which appear as small pores and which apparently allow the passage of gases but not of water.

Structure of petiole

The epidermis of the petiole is continuous with that of the stem. The parenchyma cells of the petiole, like those of the cortex, contain only few chloroplasts especially as compared with the cells of the lamina. The supporting tissues of the petiole are collenchyma and/or sclerenchyma. The vascular bundles of the petiole may be collateral, e.g. *Ligustrum* bicollateral, e.g. *Nerium* or concentric, as in certain pteridophytes and many dicotyledons. The phloem is accompanied, in many species, by groups of fibres. The arrangement of the vascular tissues in the petiole differs in different plants.

It may appear in a cross-section of the petiole as a continuous or an interrupted crescent, e.g. *Olea, Nicotiana, Nerium,* or as an entire or interrupted ring, e.g. *Ricinus, Quercus calliprinos, Q. boissieri, Citrus,* or as a ring with additional internal and external bundles, e.g. Vitis, *Platanus, Robinia, Pelargonium.* An arrangement of scattered bundles is seen in many monocotyledons and in *Rumex.* If there is a single collateral bundle in the petiole, the phloem is found on the abaxial side and if the bundles are arranged in a ring the phloem is external to the xylem on the periphery of the ring. Taking into account the variability of the vascular structure in the petiole Howard, proposed a classification relating the nodal structure at the level of the leaf gap to the vascular pattern in the leaf base and successive higher levels of the petiole or midrib.

In some plants, e.g. in *Mimosa* and *Albizzia* of the Leguminosae, the petioles or petiolules have pad-like swellings termed *pulvini* (*pulvinus*). These are capable of bringing about movement of leaves or leaflets. The pulvinus contains a large amount of parenchyma and its surface is usually wrinkled. The closure and opening movements of the leaflets may be stimulated by internal or environmental factors, and are brought about by turgor changes in the motor cells of the pulvini. In *Albizzia julibrissin* closure occurs when subepidermal cells of the dorsal side expand and those of the ventral side become compressed. Opening of the leaflets-involves the reverse changes.

The internal cells of the pulvinus surrounding the vascular tissue remain relatively unchanged during the movement of the leaflets. Electron micrographs show two features which may be connected with the movement: the presence of many vacuoles which may contain a variety of enzymes and the occurrence of thin fibrils oriented parallel to one another in the peripheral cytoplasm.

Bundle sheaths

The large veins are surrounded by much parenchyma which is poor in chloroplasts. The smaller veins, also, are usually surrounded by a layer of tightly packed parenchyma cells such a layer is termed the *bundle sheath.* In dicotyledons the cells of the bundle sheath are usually elongated in a direction parallel to the vein. Sometimes, however, for example in *Atriplex halimus,* the cells are more or less cubical. The cells of the bundle sheath are thin-walled. They may contain as many chloroplasts as the mesophyll cells, or they may contain only a few chloroplasts, or they may be devoid of them. Sometimes crystals may occur in the bundle-sheath cells.

The bundle sheaths may also surround the vein endings, e.g. in *Syringa vulgaris* and in *Atriplex halimus.* Although in most of the dicotyledons the bundle sheath consists of parenchyma cells, there are some families, such as the Winteraceae, in which the sheath is sclerenchymatous. In many dicotyledonous leaves, such as those of *Styrax officinalis* and *Quercus* spp. among others, the parenchyma of the bundle sheath extends to the epidermis on one side or on both sides of the leaf. These plates of parenchyma cells usually reach the epidermis itself, and they are termed *bundle-sheath extensions*. There is proof that the bundle-sheath extensions have a conduction function in the leaf. They conduct from the bundles to the epidermal cells.

The latter are very closely connected laterally and therefore function to conduct in a plane parallel to the leaf surface. Water may also move from the bundle sheath and its extensions through the mesophyll to the cells lining the substomatal chambers. Pizzolato *et al.* who used Wylie's method of precipitation of ferrocyanide by ferric ions in cotton leaves and examining sections in the electron microscope, found in the various cells precipitated crystals within the primary wall, in the space between the plasmalemma and cell wall, and within the protoplast. It thus seems that water flow from the tracheary elements to where it leaves the leaf takes place mainly in the apoplast.

In some plants the bundle-sheath extensions accompany the veins throughout almost their entire length, while there are other plants in which bundle-sheath extensions are completely absent. In Quercus calliprinos 94% of the total vein length is accompanied by bundle-sheath extensions; in Q. boissieri, 71%; in *Styrax* officinalis, 62%; and in Pistacia palaestina, 50%. According to Wylie, the density of the veins in mesomorphic leaves is indirectly proportional to the total length of the bundle-sheath extensions.

Bundle sheaths also occur in the monocotyledons, especially in the grasses in which two types are distinguished, one- or two-layered. The outer sheath when two layers are present, or the single sheath where there is only one, generally consists of parenchymatous cells with thin walls. The cells of this sheath contain chloroplasts, which are usually smaller than those of the mesophyll. The cells of the parenchymatous sheath may store starch and form a "starch sheath".

The cells of the inner sheath, the mestome sheath, are smaller in cross-section diameter, their walls are thickened and contain suberized lamellae. These lamellae are perforated by plasmodesmata. The mestome sheath is thus analogous to an endodermis. Suberized lamellae were also observed to occur in the walls of the parenchyma sheath of maize. The plastids in the cells of the mestom sheath, if present, are few, small and little differentiated. Single sheaths are generally characteristic of panicoid grasses and double sheaths of festucoid grasses. However, the distinction is not absolute.

Tissues of leaf

Because of its compact structure and the strength of the cuticle, the epidermis gives support to the lamina. Collenchyma is usually found close to the larger veins immediately below the epidermis and also on the leaf margins in dicotyledonous leaves. The bundle-sheath extensions may also be collenchymatous. In addition to collenchyma, sclereids are present in the mesophyll of many dicotyledons. In monocotyledonous leaves the vascular bundles are accompanied by many fibres. In the

Gramineae and in many other monocotyledons the fibres form girders on one or both sides of the bundles, and in many leaves they continue from the bundle sheaths to the epidermis, the cells of which, in such regions, may then also become fibre-like.

Secretory structures

Many of these secretory structures are of epidermal nature. The substances produced may be excreted from the cells or they may be retained only to be released upon the disintegration of the cells. Glands are present on many foliage leaves and cataphylls. These structures consist of a mass of dense parenchyma cells in which there terminates a vascular bundle. The parenchyma is covered by a glandular epidermis. The cells of such an epidermis are mostly elongated in a direction at right angles to the surface of the gland and have dense cytoplasm and large nuclei.

The presence of these glands on the petioles and on the teeth of the laminar margins, for instance, are of great taxonomic importance because of their constant position in the species, and even varieties, in which they occur. The glands on different organs or parts of an organ of the same species may secrete different substances. In *Prunus persica* and other related species, for example, the glands on the laminar teeth secrete a bitter resinous substance, while those on the petiole secrete nectar. Nectariferous glands are found on the petioles of many plants, e.g. Passiflora, Ricinus, and *Impatiens*.

In *Vicia* nectariferous tissue is present in the central portion of the stipule where it is easily distinguished because of the anthocyanins present in its cells. Another type of secretory structure is the essential oil cavity which occurs characteristically in the mesophyll of the leaves of Citrus and the secretory cavity of *Gossypium*. These are examples of lysigenous cavities in which the secretions formed in the cells are released after the disintegration of the cells, the lysis of which forms the cavity.

Secretory cavities occurring in the mesophyll of *Eucalyptus* develop lysigenously according to some authors and schizogenously according to others. Examples of other types of secretory structures found in leaves are schizogenous resin ducts, which are characteristic of the Compositae, Anacardiaceae, and Coniferales; laticifers as found in *Euphorbia;* mucilage cavities found in the Sterculiaceae, different species of the Malvaceae, Moraceae, and other families.

Secretory substances of leaves, may also be found in idioblasts. These secretory cells are classified according to the substances secreted, although cells that contain a mixture of different substances also exist. The secreting idioblasts, like other secretory structures, are of important taxonomic significance. For instance, the secretory cells of the Lauraceae, Simarubaceae, and Onagraceae contain oils. Cells containing the enzyme myrosin are found in the Cruciferae, Capparidaceae, Resedaceae, Tropaeolaceae, and Moringaceae.

Cells with a resinous content are found in the Meliaceae and in many species of the Euphorbiaceae, Rutaceae, and Rubiaceae. Tannin-containing cells are found in the Anacardiaceae, Annonaceae, Crassulaceae, Ericaceae, Euphorbiaceae, Buxaceae, Polygonaceae, Rosaceae, Tamaricaceae, and Leguminosae. Cells with tannin compounds are also found in parenchyma of fruits, e.g. of *Ceratonia.* Cells with mucilaginous contents are found in the Buxaceae, many species of the Malvaceae, Chenopodiaceae, and Rubiaceae, and in many monocotyledonous plants.

Special secretory structures are the *hydathodes* which secrete water in the form of drops from within the leaf.

This phenomenon is called *guttation.* Hydathodes secrete water which is brought to the surface by the terminal tracheids of the veins. This water passes through the intercellular spaces of the loosely packed parenchyma of the hydathode which is devoid of chloroplasts and which is called the *epithem.* The intercellular spaces open to the exterior by special pores which are of stomatal origin and which remain permanently open. The epithem may be bounded by suberized cells or cells with Casparian strips.

One of the peculiarities of gymnosperm leaves is the occurrence of *transfusion tissue.* This tissue accompanies the vascular bundles and is composed of tracheids, parenchyma, and albuminous cells. The transfusion tissue varies in amount and arrangement depending on the genera.

Initiation of growth of leaf

The main portion of the leaf primordium originates from the outermost cell layer of the shoot apex. Contrary to the situation in the grasses, in other monocotyledons, e.g. *Tulipa,* and apparently in all the dicotyledons thus far examined, the first periclinal divisions do not take place in the cells of the surface layer, but in the cells of one or more layers below it. In the apices of such plants, therefore, the surface layer does not take part in the initiation of the inner tissues of the leaf This layer enlarges by numerous anticlinal divisions of its cells and so becomes adapted to the growth of the primordium. The surface layer gives rise to the protoderm of the young leaf. In the gymnosperms there is the same amount of variation in the initiation of the leaf primordia as there is in the angiosperms. In *Taxodium distichum,* for example, the initiation of the leaf primordium originates with periclinal divisions in the cell layer immediately below the surface layer of the apex, while in most of the Coniferales and in *Zamia* the periclinal divisions take place in the surface layer of the apex as well as in the layer below it.

Most commonly the initiation of the leaf primordium commences in cell layers below the surface layer. In this case the degree to which the inner cell layers of the tunica and the neighbouring cell layers of the corpus participate in the initiation of the primordium differs, and it Is difficult to determine accurately the part played by each of them. In order to clarify this problem periclinal cytochimeras, produced by the application of colchicine, have been used. With the aid of such cytochimeras it has been possible to show from which layers of the apex the various tissues of the leaf develop. Dermen thus was able to determine that in *Vaccinium* and *Pyrus malus*, for instance, three cell layers of the apical meristem take part in the formation of the leaf. The leaf epidermis develops from the outermost layer by anticlinal divisions. The second and third layers give rise to the mesophyll and the vascular bundles.

Leaf axis

The development of the leaf axis precedes that of the lamina or of the leaflets. As a result of the rapid development the primordium becomes shaped like a gradually tapering cone the adaxial side of which is flattened. The tip of the cone functions for a while as an apical meristem, but in spermatophytes the cells at the tip of the leaf exhibit histological signs of maturation relatively soon. In certain plants, from that early stage of development when the primordium is still less than 1 mm long, all further increase in length is due to the division and elongation of cells distant from the tip of the primordium, i.e. by intercalary growth.

In leaves of ferns, however, the apical growth continues for a long period together with the addition of cells by intercalary growth in an acropetal direction. The increase in length of the axis is usually accompanied by an obvious increase in width. This thickening is particularly striking in those numerous plants in which a cambium-like region, in which the cells divide tangentially, develops along the adaxial side of the primordium.

During the early elongation and thickening of the axis of the young leaf, the cells of the adaxial margins continue to divide very frequently much -more so than the inner cells of the ground meristem. In simple leaves two wing-like strips usually develop on the margins as a result of the accelerated growth of these cells. In leaves with a petiole the marginal growth is depressed in the basal portion of the leaf axis which then develops into the petiole. In a cross-section both sides of the developing lamina can be seen to consist of protoderm enclosing a few layers of ground tissue. The new cells that are added to the different layers of the lamina originate from rows of *marginal initials* and *submarginal initials.* The marginal initials are the cells of the outermost layer on the margins of the young lamina.

Generally, in the angiosperms these initials divide only anticlinally and so add new cells to the abaxial and adaxial protoderm. In certain monocotyledons and in the bud scales of *Rhododendron spp.* periclinal divisions also occur, resulting in the addition of new cells to the nearby ground meristem.

In the variegated leaves of certain plants the white margins may develop as a result of periclinal divisions in the protoderm. In *Daphne odora* and in the hemiparasite *Eubrachion ambiguum* of the Loranthaceae the whole lamina is produced by the marginal initials only. The submarginal initials undergo divisions in various planes. These cells give rise to new cells which are added to the inner layers of the young lamina. In pinnately and palmately compound leaves the lateral leaflets develop from the adaxial marginal meristem of the axis of the young leaf as two rows of papillae.

In certain plants the order of development of the leaflets is acropetal, e.g. *Carya,* but in many other plants the order is basipetal. As is seen in *Carya,* the primordium of each lateral leaflet first develops an axis and the laminae of the leaflets later develop from the margins of these axes. The tip of the main axis develops into a terminal leaflet. The development of a lobed leaf is a result of differential activity of the leaf margin which starts early in ontogenesis. In the regions of the future lobes meristematic activity is considerable, while in the regions of future sinuses there is little or no activity.

The marginal growth apparently continues longer than does the apical growth but it, too, ceases relatively early. In *Nicotiana tabacum,* for example, Avery observed that marginal growth continues, at least in the lower part of the lamina, until that stage where the leaf is several centimetres long. In *Cercis siliquastrum* marginal growth of the lamina is completed by the time the leaf measures 2-2.5 mm. In the leaf of the Dwarf Cavendish banana, in which a marginal vein is differentiated in the early stages of development of the leaf primordium, marginal growth in the laminar part of the primordium ceases as early as between the fourth and sixth plastochron.

Marginal growth on the right side of the mid-rib ceases in the fourth plastochron when the primordium is about 5 mm long whereas, on the left side, it ceases during the sixth plastochron when the primordium, is about 20 mm, long. After the cessation of marginal growth, further growth of the lamina is brought about by cell division in the various cell layers of the lamina. These divisions are mostly anticlinal, and thus a *plate meristem* is formed. A plate meristem is one in which the planes of cell divisions in each layer are perpendicular to the surface of the organ in which the meristem occurs. The activity of such a meristem results in increase in surface area but not in thickness of the organ.

In the lamina the cells of this meristem have a stratified arrangement and therefore it is possible to trace, with relative ease, the origin of the epidermis, palisade and spongy tissues, and the vascular bundles. The regular arrangement of the cell layers is interrupted to differing extents by the development of the vascular bundles, their sheaths and supporting tissues. As a result of this, in the final stages of the expansion of the leaf surface the regular arrangement of the cell layers -becomes restricted to those areas of the lamina between the large lateral veins. The palisade parenchyma is one of the last tissues to cease growing and dividing. This tissue may continue to function as a meristem for some time after the cells of the spongy parenchyma and of the epidermis have ceased to divide.

Most of the knowledge of leaf initiation and development is based on ordinary histological observations. However, several interesting attempts were made to study the pattern of contribution of the various cell layers of the shoot apex to the leaf buttress formation and to the leaf axis and laminar growth using chimeras. The different parts of the leaf expand at different rates and in different directions. This type of growth has been termed anisotropic growth. The type of growth of a leaf is controlled by genetic factors, but it is also influenced by internal and external environmental conditions. Thus, the shape of leaves on different parts of the stem of the same plant is apparently influenced by internal factors.

Grass leaves have linear laminae and sheathing bases surrounding the stem. The development of the leaf of *Oryza sativa* as described by Kaufman will be used here as an example of the development of this type of leaf. The leaf primordia are initiated in the tunica from which the ground meristem and protoderm of the leaf develop. At the shoot apex, in the early stage of initiation, a localized protuberance appears which later becomes crescent-shaped and then as a result of further marginal and apical growth eventually surrounds the apex. As the young primordium grows upwards it becomes hood- shaped. Apical growth of the leaf ceases during the third plastochron when the primordium is about 0.9 mm long, but the margins continue to grow and the primordium elongates further. The continued marginal growth is brought about by the activity of the marginal meristem and the elongation of the primordium by a rib-meristem form of growth. A meristem of this kind is characterized by parallel series of cells in which transverse divisions take place. The rib meristem and adaxial meristem become distinguishable during the second and third plastochrons.

In the primordia of monocotyledonous leaves the processes of apical and marginal growth are simultaneous during the early plastochrons. This is contrary to the development of dicotyledonous leaves where there are two distinct stages as have been previously described. In *Oryza* the ligule is initiated during the third plastochron by periclinal divisions in the adaxial protoderm. At first the ligule is a small adaxial protuberance which, as a result of continued periclinal divisions in the protoderm, expands laterally towards the margin of the sheath. The auricular primordia are apparently derived from both the sheath and ligule protoderm as well as from the ground meristem. Continued elongation of the lamina and sheath, at this stage, still results from the activity of the rib meristem, whereas the extension of the lamina wings is caused by the activities of the marginal and of the distinct plate meristems.The development of the sheath differs from that of the lamina in that no distinct plate meristem is seen in the wings of the sheath and the extension of them is accomplished, primarily, by the activity of the marginal meristem and by the enlargement of the cells derived from it.

As the differentiation in the lamina precedes that in the sheath, the meristematic activity becomes more and more restricted to the base of the sheath where the region of actively dividing and enlarging cells should, therefore, be regarded as an intercalary meristem. The direction of the cellular differentiation and maturation in the leaf of Oryza and *Musa,* for instance, is basipetal. The development of the laminar mesophyll in the former is depicted.

Monocotyledonous leaves borne vertically, without distinction of adaxial and abaxial sides, are called *unifacial* leaves. This type of leaf was thought to develop on a sympodial pattern. The primary apical meristem of the primordium was considered to be responsible for initial radial growth. It ceases its activity very early and a new centre of growth arises on the abaxial side of the primary leaf apex. This secondary apex was held to be responsible for the longitudinal extension of the leaf. The unifacial leaf portion, above the sheathing base, which develops from the secondary apex may be cylindrical, as, for instance, in *Allium cepa* and *Juncus* maritimus, or it may be flattened laterally as in Iris. The radial orientation of this ensiform or sword-shaped leaf is a result of emphasized adaxial meristematic activity and suppressed marginal meristematic activity. Adaxial meristematic activity is expressed very early in development of the leaf and causes the ventral surface of the primordium to protrude considerably. This protuberance was in the past misinterpreted, as stated by Kaplan, as a primary leaf apex. Apical growth in Acorus ceases early in ontogeny and subsequent growth in length is largely basal and intercalary. The radial growth of the laterally flattened portion of the leaf is first a result of the emphasized activity of the adaxial meristem and subsequent intercalary meristematic activity on both sides of a secondary midrib. This Midrib develops early along the centre of the leaf axis, and procambium differentiates in it. Then procambial strands are initiated in pairs, in both sides of the primordium, from derivatives of intercalary meristem in the adaxial and abaxial wing of the leaf. Procambial differentiation is continuous and acropetal.

The radial type of unifacial blade occurring in many monocotyledons has, according to Kaplan, evolved by a divergent course of lamina morphogenesis in which growth in volume dominates over surface expansion. Kaplan's view of the Acacia phyllode, is that this organ is homologous with the rachis of the bipinnate leaf plus a small sector of the petiole and is thus a unifacial leaf form of dicotyledons. The metamorphosis of the phyllode from a bipinnate type leaf does not involve distal laminar suppression and compensatory petiolar elaboration, but rather takes place by an alternative mode of morphogenesis. Instead of the developing blade forming a series of adaxially situated pinna primordia, a single adaxial meristem is responsible for the vertical orientation of the phyllode-type of blade.

Leaf abscission

Leaves, floral parts, fruits, and often branches, e.g. *Taxodium. Populus, Naucleopsis,* and *Perebea* may be shed. In some species, e.g. in the tumbleweed *Kochia indica,* the whole aerial part of the plant separates from the root. Leaves of gymnosperms and woody dicotyledons are usually shed as a result of changes that take place in the tissues of the leaf base prior to leaf death. In the base of mature deciduous leaves a narrow zone, the so-called *abscission Zone* or *region* can be seen. This zone can be distinguished histologically from the surrounding tissues and externally its location is marked by the presence of a shallow groove or by a difference in the colour of the epidermis. The vascular system in this region is usually concentrated in the centre and the sclerenchyma and collenchyma are less well developed or absent.

In the abscission zone two layers are discernible-an *abscission* or *separation layer* through which the detachment of the organ occurs and a *protective layer* which protects from desiccation and from the entry of parasites to the exposed surface. In many abscissing organs the preparation of the separation layer takes place during normal ontogeny, but may also occur a short time before shedding of the organ in response to conditions causing abscission. Several histological processes were observed during abscission, not all of them in the same species.

In the abscission zone lignification of the parenchyma cells, tylose formation in the tracheary elements, and callose deposition in sieve elements was also observed. The final event in abscission involves the enzymatic degradation of cell walls. This starts with removal of calcium and pectin from the middle lamellae and continues with the hydrolysis of the cellulose walls. At the end, the sclerified tracheary elements rupture. According to Facey, the middle lamellae between the cells of the vascular tissue also disintegrate. Xylem vessels and interstitial parenchyma of vascular bundles may also become weakened by hydrolysis of cell-wall material prior to separation. Golgi vesicles may be involved in secretion of abscission-specific enzymes.

In some species no dissolution takes place and abscission is apparently effected by physical stresses. The latter is the case in most monocotyledons and herbaceous dicotyledons. After the leaf dies it, base breaks away from the stem through the weakened region of the abscission zone. The *protective layer* may be of two types—primary or secondary, i.e. periderm. The primary protective layer-is formed as a result of the lignification and suberization of the parenchyma cells in this region or of the cells arising from them by irregular cell divisions. An opinion exists that the substance appearing in the cell walls of the protective tissue, and which has been defined as lignin, is really wound gum which gives reactions similar to lignin. The time at which the primary protective layer and periderm appear differs in different plants.

Chapter 6

Anatomy of Root

Roots bear no appendages that are comparable to the leaves of the stem: roots are devoid of stomata and their branches originate in the relatively mature tissue of the pericycle in contrast to the stem where the branches originate from the apical meristem. Roots also possess a root cap which has no parallel in stems. Much variability exists in the shape and structure of roots. Environmental conditions of often influence the root system.

On the basis of origin two types of roots—primary roots and adventitious roots—are distinguished. Primary roots develop from the apex of the embryo that is destined, from its origin, to give rise to a root, and from the pericycle of relatively mature parts of roots, while adventitious roots develop from other tissues of mature roots or from other parts of the plant body, such as stems and leaves.

In most dicotyledons and gymnosperms the root system consists of a tap root from which side branches arise. The order of appearance of the lateral roots is from the root neck towards the root tip, but in some cases the primordia of some of the lateral roots remain dormant. The mature portions of the root, which usually undergo secondary thickening, function only as a holdfast in the soil and to store reserve materials. The uptake of water and salts is carried out mainly by the extremities of the root system which are still in the process of primary growth. The roots of mature monocotyledons are usually adventitious and develop from the stem. They may branch several times, as do the roots of dicotyledons, or they may be unbranched. Generally such roots do not develop secondary thickening.

In the gymnosperms and dicotyledons this root develops into the tap root with its branches. In monocotyledons this root usually dries out early in the growth of the plant and the root system of the mature plant consists of numerous adventitious roots. The apical meristems of lateral roots develop deep within the inner tissues in contrast to the buds of the shoot which develop from outer tissues. Therefore the branching of roots is *endogenous.*

Root cap

Root cap is situated at the tip of roots, it protects the root pro-meristem and aids the penetration of the growing root into the soil. The root cap consists of living parenchyma cells which often contain starch. These cells may have no special arrangement or they may be arranged in radiating rows which originate from the initials. In many plants the central cells of the root cap form a more distinct and constant structure which is termed the *columella.* Root cap cells secrete a polysaccharide slime. It was found that the process of secretion is accompanied by hypertrophy of the dictyosome cisternae forming large vesicles. The vesicle contents are subsequently released from the protoplast by fusion of the membranes of the vesicles with the plasmalemma. The secretion moves outwards through the wall. It has long been known that the root cap controls the geotropic growth of the root.

Scientists of the last century had already demonstrated that removal of the extreme tips of roots prevents their geotropic response. Later it was suggested that certain solid cell inclusions, termed *statoliths*, principally starch grains, occurring in the root cap cells transmit gravitational stimuli to the plasmalemma of the statolith-containing cells *(statocysts* or *statocytes).* Starch grains were, indeed, seen to collect at the lower side of the root cap cells.

Electron microscope studies of root cap cells of *Vicia faba* roots subjected to geotropic stimulation showed that the amyloplasts sedimented to the lower side of the cells, thus displacing the ER and other organelles to the upper halves of the cells. The existence of two highly mobile components, i.e. amyloplasts and ER, in statocytes is also supported by experiments carried out by Juniper and French.

However, Sievers and Volkmann and Volkmann who worked on roots of *Lepidium sativum* and some other plants, suggested that geoperception in roots may be a function of pressure extended differentially by amyloplasts on static ER complexes. The site of geoperception, the root tip, is separated from the site of gravity reaction, the elongating zone of the root.

The nature of the stimulus transmitted from the root tip to the zone of cell elongation is as yet not known but there is some evidence that it may be an unequal distribution of abscissic acid. The root cap develops continuously. The outermost cells die, become separated from one another, and disintegrate, and they are replaced by new cells which are produced by the initials. The rate of production and sloughing of the cap cells is high.

Epidermis

Epidermal cells of roots are thin-walled and are usually devoid of cuticle, although sometimes the outermost cell walls, including those of the root hairs, undergo cutinization. On those parts of roots that are exposed to the air and on those parts in the soil on which the epidermis persists for a longtime, the outermost cell walls become thick, and may sometimes contain lignin or dark-coloured substances which have not been fully identified. The region of root hairs is usually restricted to one or a few centimetres from the root apex root hairs are absent close to the apical meristem and they usually die and dry out on the more mature portions of the root.

Certain herbaceous plants, and especially water plants, lack root hairs. Plants that usually grow in soil and which produce root hairs fail to do so when they are grown in water. Calcium is one of the factors controlling the normal development of root hairs. In some plants the root hairs remain on the root for a long time. In *Gleditschia triacanthos*, for example, the root hairs remain viable for some months and their walls become thickened.

Long-lived root hairs have been found in certain species of the Compositae and in some plants of other families but it is doubtful whether these root hairs take part in the uptake of water from the soil. Epidermal cells of young roots and root hairs of some genera secrete fibrillar mucilage forming a mucilaginous layer. In plants layer was found to be colonized by bacteria. In *Sorghum,* in addition to fibrillar 'mucilage, drops of mucilaginous character are secreted near the tip of the root hairs.

Exodermis

The exodermis is similar in structure and cytochemical characteristics to the endodermis. An almost continuous suberin lamella lines the primary cell wall internally, and it in turn is usually lined with layers of cellulose which develop centripetally. Also lignin is often deposited in the walls of these cells and, in certain cases, Casparian strips have also been distinguished. The cells of the exodermis contain viable protoplasts even when mature. In the pteridophytes no exodermis is developed. Sometimes, however, for instance in *Ophioglossum*, certain fatty substances are deposited in the walls of the subepidermal cells, but no special suberin lamella is formed.

The thickness of the exodermis varies from a single cell layer to many layers. The exodermis may sometimes be accompanied, on its inner side, by sclerenchyma as, for example, in the root of *Ananas*, of some grasses and Cyperaceae. In *Phoenix* the exodermis is fibrous. The endodermis consists of a uniseriate cylinder of cells and develops, apart from a few exceptions, in all vascular plants. This layer of cells represents the inner boundary of the root cortex.

Because of its physiological and phylogenetic importance, the endodermis has been very thoroughly investigated. In that part of the root where the primary vascular system is starting to mature, Casparian strips appear in the radial and cross walls of the endodermal cells. These strips are not merely wall thickenings but integral parts of the primary wall and the middle lamella in which suberin and lignin are deposited.

In the electron microscope the Casparian strip appears as a slightly thickened, homogeneous electron opaque wall region. If plasmolysis is brought about in endodermal cells the protoplast withdraws from the tangential walls but remains attached to the Casparian strips. This type of plasmolysis has been termed band plasmolysis. Plasmodesmata occur in all wall portions of the endodermal cells except for the Casparian strips -to which the plasmalemma adheres. The Casparian strips prevent inward flow of water and nutrients through the apoplast. During the primary development of the root the endodermal cells are capable of much additional growth.

This feature is striking during the endogenous initiation of lateral roots from the pericycle which is accompanied by the division and stretching of the neighbouring endodermal cells.

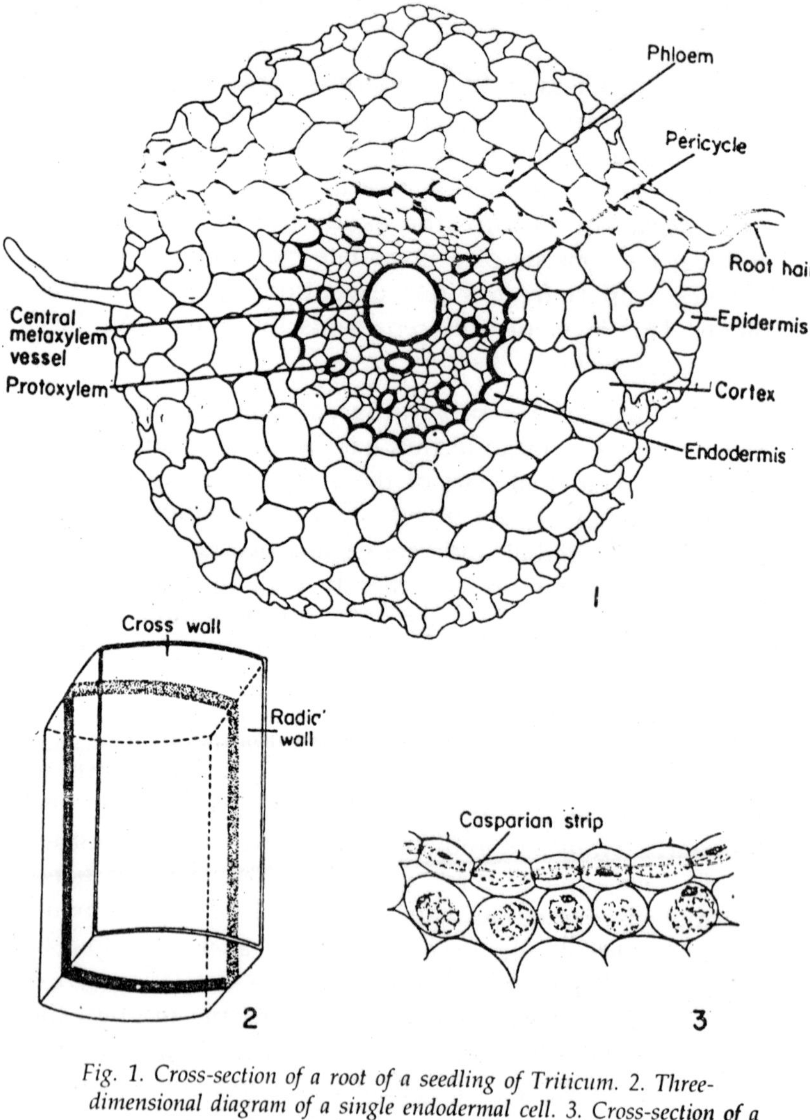

Fig. 1. Cross-section of a root of a seedling of Triticum. 2. Three-dimensional diagram of a single endodermal cell. 3. Cross-section of a root showing part of the endodermis in a state of plasmolysis.

In addition, the endodermal cells in certain roots continue to divide anticlinally during the early stages of secondary thickening. Casparian strips develop in many of the cells thus formed. In many of the angiosperms, pteridophytes, and some gymnosperms, the endodermis remains in the primary form and is shed together with the cortex With the development of secondary thickening and periderm. However, in other angiosperms in which there is no secondary thickening, an almost continuous lamella of suberin develops oil the inner side of the entire primary wall, including the Casparian strips. This lamella characterizes the second-stage of development of the wall and in the third stage a layer of cellulose is laid down centripetally on the inside of the suberin lamella. This layer may reach a very considerable thickness on the radial walls, the walls parallel to a cross-section of the root, and on the inner tangential walls of the endodermal cells. This type of endodermal cell is common in the roots of most monocotyledons. These thickened endodermal cell walls may become lignified.

The endodermis of the conifers is characterized by the second stage of wall development only, i.e. by the development of a suberin lamella on the inner side of the walls. The additional wall layers of the endodermal cells do not develop simultaneously in all the endodermal cells as seen in a single cross-section. Casparian strips and the successive stages of the development of the typical wall first appear opposite the phloem strands from where the development spreads towards those endodermal cells that are opposite the xylem strands. Because of the delay in wall differentiation of the endodermal cells opposite xylem these cells often have Casparian strips only. These cells are termed passage cells as they are thought to provide passage for substances between the cortex and vascular cylinder. The passage cells may remain unaltered throughout the entire life of the root or they, too, may develop thick secondary walls as do the other endodermal cells. The production of the suberin lamellae on the endodermal cell walls results from the polymerization of unsaturated fatty compounds which is brought about by oxidases and peroxidases. The peroxidases are brought to the endodermal cells via the sieve elements. This has led van Fleet to suggest that this is the reason why the greatest amount of suberin is laid down on the inner walls of the endodermal cells and why the passage cells, which lack suberin, appear mostly opposite the xylem, and not the phloem, strands.

Vascular cylinder

In root, the vascular cylinder occupies the central portion of the root. It is more clearly delimited from the cortex in roots than in the stem, because of the presence of the endodermis which is characteristically better developed in roots. The primary vascular tissue is surrounded by a region of cells which is termed the pericycle. The pericycle generally consists of one or more layers of thin-walled parenchyma cells. It is in direct contact with the protophloem and protoxylem and can already be distinguished prior to the lignification of the protoxylem elements. The pericycle retains its meristematic characteristics. The primordia of the lateral roots in all spermatophytes and the phellogen and portions of the vascular cambium in the dicotyledons develop from the pericycle. In monocotyledons the phellogen usually develops in the outer parts of the cortex.

In roots of many Gramineae and Cyperaceae the outermost elements of the protoxylem may develop in the regions of the pericycle and, in the Potamogetonaceae, even the phloem elements may do so. In such cases the pericycle is not continuous. In monocotyledons, where there is usually no secondary thickening, sclerification takes place in part or all of the pericycle.

The phloem strands are always separate and they are concentrated on the periphery of the vascular cylinder. The xylem strands may be in separate units on the periphery of the vascular cylinder or they may extend into the centre and then, as in many plants, the xylem appears star-shaped in cross-section. This structure led many workers to regard the vascular cylinder of the root as being a protostele. In many plants, and especially in the monocotyledons, the xylem strands do not reach the centre of the vascular cylinder which is then occupied by a pith. The tracheary elements in the root mature centripetally and therefore the xylem is exarch, i.e. the protoxylem is situated on the outer side of the metaxylem.

The differentiation of the phloem is also centripetal so that the protophloem is closest to the pericycle, while the metaphloem is closest to the axis of the root. The number of protoxylem groups in the root, i.e. whether one, two, three, etc., is expressed by the terms *monarch, diarch, triarch* respectively, and a root in which there are many protoxylem groups is said to be *polyarch.* Diarch roots are found, for example, in *Lycopersicon, Nicotiana, Beta, Raphanus, Daucus,* and *Linum.* A correlation exists between the diameter of the vascular cylinder and the number of protoxylem groups and the presence or absence of a pith. When the diameter of the vascular cylinder is large, a pith is usually present and the number of protoxylem groups is large. Variations in these features may be found even within the same plant. In gymnosperms and dicotyledons the number of the xylem strands in the root is generally small. In these groups of plants the roots are usually di-, tri-, or tetrarch but there are some dicotyledonous species in which there are more xylem strands. The water-plant, *Trapa natans,* has a thin root which is monarch. In monocotyledons the number of xylem strands in the seminal roots is small, as in dicotyledonous roots, but the adventitious roots are polyarch and the number of strands in the Palmae and the Pandanaceae may be 100 or more.

In most plants there are interconnections between the different tracheary elements which appear separate in cross-section of the root. In roots with a pith it is also possible to find, here and there, lateral connections between the phloem groups. However, there are plants that have no lateral connections between the phloem strands or the different tracheary elements. The primary phloem of the roots of most plants does not contain fibres but in certain plants, such as those of the Papilionaceae, Malvaceae, and Annonaceae, fibres are found in the primary phloem. In many mature roots that do not have secondary thickening the parenchyma associated with the primary vascular tissues becomes sclerified. In many Coniferales, except for the Taxodiaceae, Cupressaceae, and Taxaceae, resin ducts are present in the region of the primary conducting tissues.

Mycorrhizae

The epidermis and cortex of roots of many plants are often associated with soil fungi. This widespread association between fungal hyphae and the young roots of higher plants is known as *mycorrhiza* (Greek *mykes* mushroom, *rhiza* root). Usually this is a symbiosis: both the higher plant and the fungus derive benefit from the association.Two main types of mycorrhizae are recognized: *ectomycorrhizae* and *endomycorrhizae.* In the ectomycorrhizae the fungus produces a mycelium on the root surface. Hyphal strands penetrate the root between the cortical cells and form there a net. In the endomycorrhizae the fungus forms an inconspicuous mycelium on the root surface but invades the interior of the root cells. Ectomycorrhizal roots are short, branched, often dichotomous and appear thicker than the normal roots.

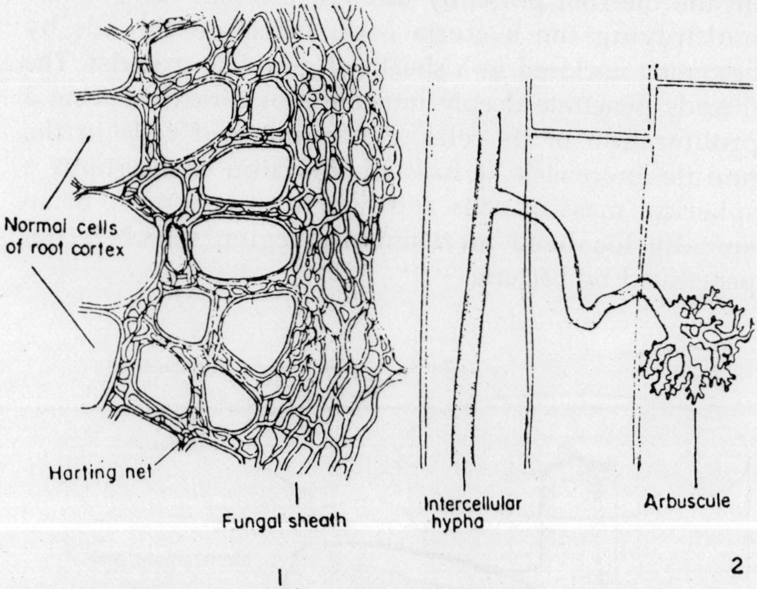

Fig. 2. Diagrmas of mycorrhizae. (1) Ectomycorrhiza based on a micrograph of Meyer. (2) Endomycorrhiza based on a micrograph of Cox and sanders.

Certain mycorrhizae have been called ectoendomycorrhizae because of the occurrence of penetration hyphae of the ectomycorrhizae type into the host cells. Electron microscopical studies of endomycorrhiza roots showed that the plasmalemma of the host cells surrounds the individual branches of the fungal arbuscule. Mycorrhizae occur on all types of plants: ferns, wheat, orchids, and forest trees. They are very often essential for successful plant growth. The role of the mycorrhizal fungus appears to be the absorption of nutrients from the soil.

Root nodules

Root nodules develop as a result of penetration of nitrogen-fixing bacteria into the root cortex. The bacteria invade the root primarily through the root hairs. While multiplying the bacteria form infection threads by becoming enclosed in a sheath of gum-like material. The threads penetrate deeply into the root cortex inducing a proliferation of its cells. The number of cells in the nodule increases at first by division throughout a spherical mass of cells and later by the activity of an apically localized meristematic region which is not penetrated by bacteria.

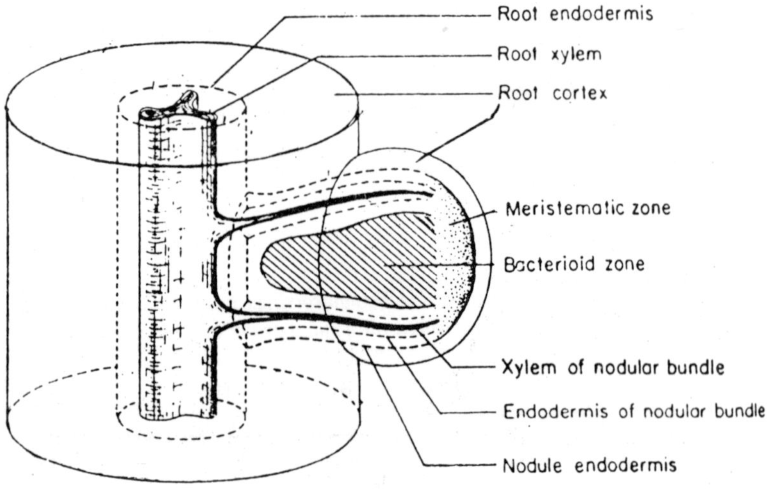

Fig. 3. Diagramatic representation of leguminous root and nodule.

The differentiated cells of the inner region, the bacterioid zone, contains bacteria released from the infection threads. The nodule at this stage resembles superficially a primordium of a lateral root. The root epidermis is broken as the nodule enlarges but the nodule does not emerge from the cortex. The cells of the root cortex divide and stretch considerably so that they remain as the outermost layer of the nodule. Branches developing from the root vascular tissue surround the bacterioid zone. Each bundle has a parenchymatic sheath and an endodermis. In some plants, the sheath cells of these bundles develop wall protuberances characteristic of transfer cells serving in short distance transport.

secondary growth in roots

There is great variation in the secondary growth in different roots. The tap-root and main lateral roots of gymnosperms and woody dicotyledons usually have secondary thickening, but the smallest branches 'do not. In the roots of some herbaceous dicotyledons secondary thickening may be completely absent, vestigial (e.g. *Ranunculus)* or it may be well developed (e.g. *Medicago).* Roots of most monocotyledons are devoid of secondary thickening.

After these cambial cells have produced a few secondary elements the pericycle cells on the outer sides of the protoxylem groups begin to divide, and the inner cells resulting from these divisions form cambial cells. These strands of cambium unite with those on the inner sides of the primary phloem strands.

At first the cambium has an undulating shape, as seen in cross-section of the root; but as the development of the secondary xylem on the inner side of the phloem strands precedes that of the secondary xylem external to the protoxylem groups, the cambium soon becomes circular in cross-section.

The secondary tissues of the root differ to some extent from those of the stem. The root may have a higher bark to wood ratio; relatively less fibres in the bark; less fibres in the wood; larger and more uniform vessels; less distinct growth increments; a larger volume of ray tissue and wider and longer tracheids in the gymnosperms, often with multiseriate arrangement of pits; a relatively large volume of living cells; less tannic substances.

Lateral roots

Lateral root primordia seem to occur in many plants in a more or less regular sequence. It was found that the smaller the number of protoxylem poles the greater the degree of regularity in the arrangement of the lateral roots. It seems that the spacing of lateral root primordia in the horizontal plane is determined by their relationship to the developing vascular system, and that little or no inhibition or competition exists between primordia situated opposite adjacent protoxylem poles. In longitudinal plane, several factors seem to control the spacing of lateral root primordia. These are the effect of older primordia on the future sites of lateral root initiation in the same sector, and the effect of parent root apex.

Initiation of lateral roots usually commences immediately behind the region of root hairs; but in certain plants, especially water plants, e.g. *Eichhornia,* the lateral roots begin to form, in the young pericycle, below this region. In, gymnosperms and angiosperms the lateral roots are commonly initiated in the pericycle whence they pass through the cortex of the parent root to the exterior. In pteridophytes the lateral roots mostly are initiated in the endodermis. A definite relationship exists between the position of the protoxylem groups and phloem strands and the position of the initiation of lateral roots.

In triarch roots or those with more than three protoxylem groups the lateral roots usually develop opposite the protoxylem groups. But in certain plants, for instance species of the Gramineae, Cyperaceae, and Juncaceae, the lateral roots develop opposite the phloem strands. In diarch roots the primordia of the lateral roots appear opposite the phloem strands or close to the protoxylem groups, and so two rows of lateral roots develop. In some diarch roots, lateral roots may develop on both sides of the protoxylem groups so that, in such cases, four rows of lateral roots develop. In the fleshy carrot root additional lateral roots develop at the base of earlier formed lateral roots that have dried out. The initiating divisions are periclinal. Bell and McCully have observed that in Zea mays the early stages of lateral root initiation, adjacent to a protoxylem pole, involve not only pericycle cells but also a considerable number of parenchyma cells of the adjacent portion of the stele.

As a result of further growth the primordium penetrates through the cortex of the parent root. It is possible to distinguish the zones of primary tissues, apical meristem and root cap of the lateral root primordium even before it appears on the surface of the parent root. Different opinions exist as to how the passage of the growing lateral root is effected through the cortex of the original root. According to one view the lateral roots partially digest the cortical tissue during penetration, while according to another view the process of penetration is purely a mechanical one. However, it is generally agreed that the developing lateral roots do not form and connection with the tissues through which they penetrate.

As the lateral roots originate in the pericycle the distance between the two vascular systems is small. Of the intervening cells, which also develop from the pericycle, some differentiate into sieve elements and some into tracheary elements. The xylem of the lateral roots of many monocotyledons is connected with two or more xylem strands of the original root. This can be seen in *Monstera*, for example, where the connections are not only with the peripheral xylem strands but also with the innermost large vessels of the metaxylem. This is brought about by the modification, into tracheary elements, of parenchyma cells between the xylem and phloem strands.

Adventitious roots

Adventitious roots may develop from large roots, from the hypocotyl of young plants, from the primary and secondary body of stems, and from leaves. In the roots and stems of most plants adventitious roots develop endogenously, but there are examples in which the development is exogenous.

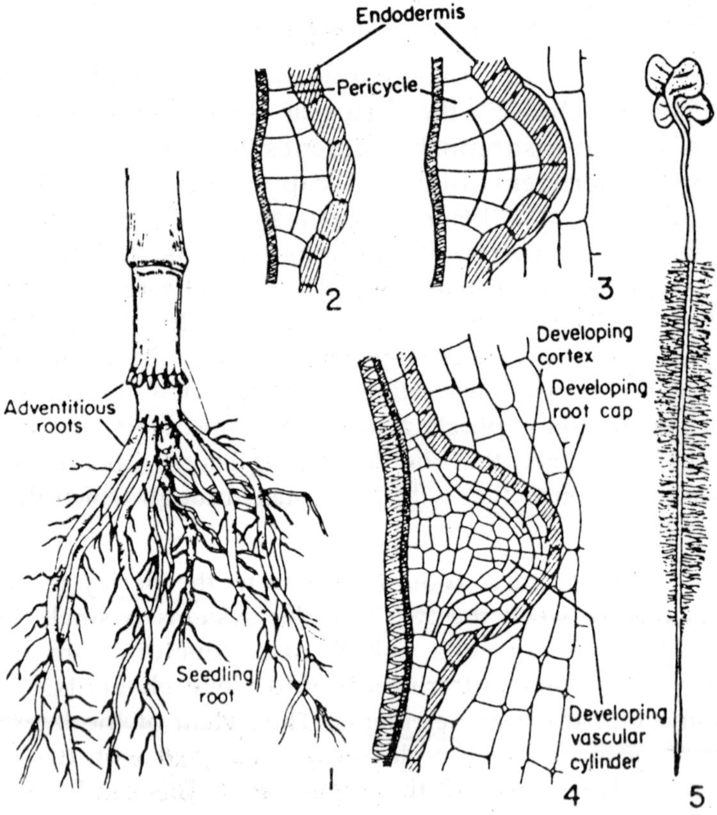

Fig. 4. (1) Basal portion of Zea mays with seedling roots and the adventitious roots. (2)-(4) Portions of radial longitudinal sections of roots of Hypericum showing early stages in the development of lateral roots. (5) Seedling of Sinapis alba with root hairs.

Primordia of adventitious roots may be formed by the following tissues: the epidermis, together with cortical tissue, of buds and hypocotyls (e.g. *Cardamine pratensis, Rorippa* austriaca); stem pericycle (e.g. *Coleus,* Zea mays); ray parenchyma between pericycle and cambium (e.g. Tropaeolum majus, Lonicera japonica, Tamarix); phloem ray parenchyma (e.g. *Hedera helix);* non-differentiated secondary phloem and cambium between the vascular bundles (e.g. *Rosa);* interfascicular cambium and pericycle (e.g. *Portulaca oleracea); interfascicular cambium,* pericycle, and phloem e.g. *Begonia); the* pith of the stem (e.g. Portulaca *oleracea); par*enchymatous interruptions in the secondary xylem which are formed due to the presence of leaf gaps (e.g. *Ribes* nigrum) or buds (e.g. *Cotoneaster dammeri); tissues* of leaf margins and petioles.

Auxins are often used to promote adventitious root formation. Gramberg, has observed, in leaf petioles of *Phaseolus vulgaris* treated with IAA, that the first symptom of adventitious root induction is the formation of new protein and nucleic acids in the nucleus. The development of adventitious roots has been described in some species of *Salix.* In these species the adventitious roots develop from primordia which appear in the stem prior to its removal as a cutting. These primordia are formed from secondary' parenchymatous tissue in the leaf or branch gaps. Several layers of cells external to the cambium take part in the formation of a primordium, to whose inner side cells are also added by the cambium. The primordium becomes dome-shaped as a result of the intensified growth of the secondary xylem immediately internal to it. These primordia remain dormant within the inner bark as long as the branch is not removed from the tree. The differentiation of these primordia is extremely slow so that even in 9-year-old branches, the typical root tip structure is not discernible.

After the first year of growth, additional primordia may develop vertically above and below the first-formed primordia on both branches left on the tree and on cuttings. On cuttings most of the primordia develop rapidly into roots. Similar adventitious root primordia have been observed on woody roots of *Zygophyllum dumosum.* Plant species differ from one another in their ability to produce roots on cuttings.

Cuttings of plants with dormant or preformed adventitious root primordia (e.g. *Salix)* root easily as do many plants with broad vascular rays but without such primordia. The vascular tissues of the root are attached directly to the stem vascular bundles, often penetrating deeply into the stem. An extensive peripheral attachment via girdling vascular bundles is also present. The root vessels break up into aggregates or clusters of narrow elements which run parallel to the stem bundles. The distal elements of these clusters are applied to the metaxylem vessels of the stem bundles. The terminal clusters of the root xylem frequently spread around the stem vessels, thus making very wide contacts. The metaxylem of the stem bundles is not usually modified.

Chapter 7

Anatomy of Flower

Flower are the reproductory organs of a plant. They consists of an axis on which the rest of the floral organs are borne. That part of the axis that represents the internode terminated by the flower is termed the *pedicel.* The distal end of the pedicel is swollen to various extents and this portion is termed the floral *receptacle* or *thalamus*. The floral organs are attached to the receptacle. A typical flower has four types of organs. The outer most organs of the flower are the sepals which together constitute the calyx which is usually green and is found lowest on the receptacle. On the inside of the sepals is the *corolla,* consisting of the *petals* which are generally coloured. These two types of organs together form the *perianth,* however, sometimes one of them may be lacking. When all the organs of the perianth are similar they are termed *tepals*. Within the perianth two kinds of reproductive organs are found: externally the *stamens* which together form the *androecium,* and internally the carpels which form the *gynoecium.* The arrangement of the floral organs on the receptacle may be spiral or whorled, and both types of arrangement may occur in the same flower.

In most flowers in which the arrangement is whorled, the organs of each whorl alternate with those of the neighbouring whorl. The floral organs may be free or fused. Fusion of organs of the same type is termed *cohesion,* and that of different types of organs, *adnation*.

The term *pistil* is no longer used as it is not sufficiently well defined. This term was used both for each of the free carpels of a flower as well as for the unit which is formed by the fusion of a few carpels. Therefore, in this book we shall refer to all the carpels of a flower, whether they are free or fused, as the gynoecium. The stamen consists of a *filament* which distally bears the *anther. Two* lobes can be distinguished in the anther and they are attached to a continuation of the filament which is termed the *connective*. Each of these lobes contains two *pollen sacs* in which *pollen grains* are found. The pollen grains in their early stage of development are the microspores of the angiosperms. The free carpel or fused gynoecium usually consists of the following three parts: the *ovary*, a hollow body which contains one or more *ovules*: the *style* which results from the elongation of the ovary wall; and the *stigma* which is that part at the top of the style which has a surface structure enabling pollination. The ovules are attached to a special thickened region of the carpel wall which is termed the *placenta*. When the carpels are found on the highest level of the floral axis the ovary is termed superior and the flower *hypogynous*.

In certain plants the perianth and stamens are located on the edge of a laterally expanded disc which raises them above the ovary; such a flower is termed *perigynous*, and its ovary is said to be *intermediate* or *pseudo-inferior*. The concave disc may completely enclose the ovary so that it is then found below the other floral organs; in such a flower the ovary is said to be *inferior*, and the flower *epigynous*. In some plants, e.g. *Capparis*, only that part of the receptacle that bears carpels is elongated; such a structure is termed a *gynophore*. An elongation of that part of the receptacle that bears the carpels and stamens, e.g. *Passiflora, is* termed *androgynophore*.

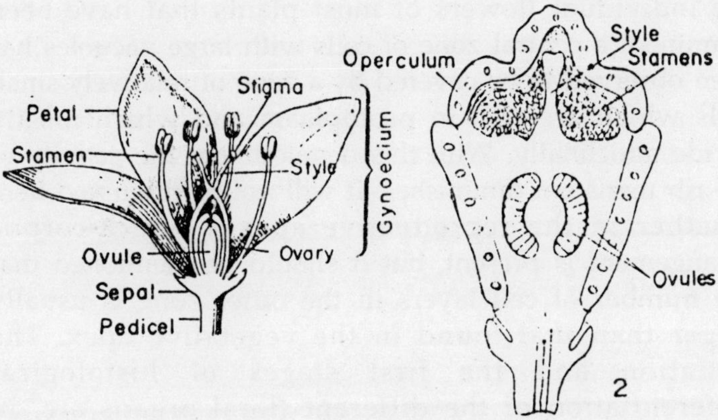

Fig. 1. (1) A flower of Fagopyrum sectioned longitudinally. (2) A flower bud of Eucalyptus globulus sectioned longitudinally.

The morphological changes that take place are usually rapid and obvious. In many plants the axis elongates abruptly during this time or immediately afterwards and the apex widens and often becomes slightly flattened. The sudden elongation of the flower-bearing axis prior to flowering is especially noticeable in those plants in which the internodes are very short and in which the foliage leaves form basal rosettes, e.g. in many species of the Compositae, in the Gramineae, and in the banana.

The elongated axis may bear a single flower but usually it bears an inflorescence. The terminal meristem and the meristems in the axils of the bracts may develop flowers in the order that reflects the type of inflorescence. In a vegetative apex the apical meristem usually continues to function above the developing leaf primordia, but in the case of a developing flower the meristematic zone in the apex diminishes until it is lost completely, or until a small, non-active residue alone remains.

In the apical meristem of the inflorescences and .of the individual flowers of most plants that have been examined, a central zone of cells with large vacuoles has been observed to be covered by a zone of relatively small cells which are rich in protoplasm and which mostly divide anticlinally. With this development the activity of the rib meristem diminishes. It will not be discussed here whether in the reproductive apex a tunica-corpus arrangement is present, but it should be mentioned that the number of cell layers in the outer zone is usually larger than that found in the vegetative apex. The initiation and the first stages of histological differentiation of the different floral organs are, in principle, similar to the early ontogenetic stages of bracts and foliage leaves. The morphological and functional differences between the different floral organs, are apparently related to a series of physiological processes which take place during the different stages of floral differentiation. This assumption is supported by the results of experiments which involved the incising and dissecting of the primordia of floral organs at different developmental stages. The developing floral organs usually appear in a distinct acropetal order, i.e., the youngest organs are closest to the apex.

In zygomorphic flowers unequal development of the floral organs is already obvious in early ontogenetical stages. In *Reseda,* for example, the adaxial side of the receptacle is already better developed than the abaxial side in the earliest stage of development, and the abaxial petal appears only after the adaxial stamens are almost fully developed. In those flowers in which certain organs are partially reduced, as, for instance, in the case of the sterility of some of the stamens, the development of these organs is retarded in relation to that of the other normal organs. The development of the reduced organs ceases completely before mature shape is obtained. The female flower of *Musa* may be given as an example of such development. In certain species the appearance of primordia of reduced organs may be retarded in relation to that of the similar normal organs. Examples also exist in which the carpels mature before the stamens. This feature is apparently brought about by a local concentration of growth substances.

In flowers in which the gynoecium is *apocarpous*, i.e. in which the carpels are free, each carpel primordium appears in its earliest stage of development as a rounded buttress which is similar to the primordia of the other floral organs and leaves. At a later stage of development the carpel primordium resembles that of a peltate leaf. Still later a depression appears at the tip of the primordium and as a result of unequal development, which then commences, an abaxial lip, from which the dorsal side of the carpel develops, is formed. The adaxial lip of the primordium develops more slowly and forms the "sill" of the carpel. This sill may enlarge and form the margins of the carpels. It is assumed that the sill originally consisted of the two basal laminar lobes which, as in the ontogeny of the peltate leaf, have fused. The area where the two margins fuse is termed the *cross-zone*. The ovule or ovules develop from this zone. In most cases the dorsal side of the carpel folds and closes over the sill. In a few genera of the monocotyledons, e.g. plants belonging to the Butomaceae, the primordia of the carpels are horseshoe-shaped, in cross-section, and remain so almost until the maturation of the carpels, and so no sill is developed. In this case the fusion of the margins of the carpels is incomplete even in the mature carpel. In flowers with *syncarpous* gynoecia, i.e. those in which the carpels are fused, the ovaries may develop in two ways. In the first type, carpel primordia first appear separate and fuse later as a result of lateral growth (ontogenetic fusion). Subsequently, the carpels grow upwards as a unit on which the primordial apices are borne. In the other type of development the carpels are already joined in the earliest stages of primordial development (congenitally fused) so that the ovary wall rises, from its initiation, as a ring. The regions of fusion between the carpels are distinguishable by inwardly directed folds.

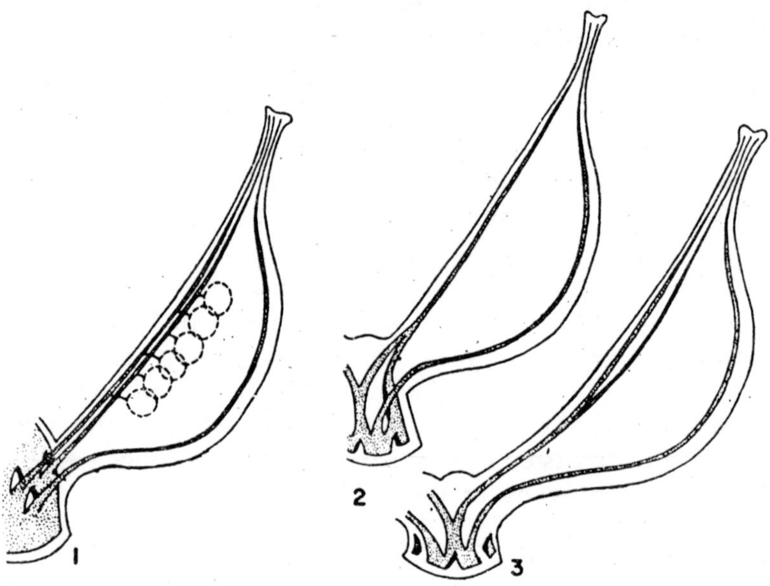

Fig. 2. Diagrams illustrating the carpellary vascular supply

Fusion of flower organs, whether postgenital or congenital, presents many developmental problems. During the very early stages of development of an inferior ovary, a depression can be distinguished in the centre of the developing flower. This depression gradually deepens with the development and growth of the flower organs. In the flower of Musa, three separate rounded carpel primordia can be distinguished before the formation of this depression. In *Downingia*

(Campanulaceae), as described by Kaplan, this depression is a result of the growth of the common bases of the developing floral organs. Firstly, the common bases of the sepal lobes grow to form a floral cup. The initiation of the following floral whorls occurs along the inner margins of the cup. Continued basal growth of the cup-shaped bud results in the formation of the elongated inferior ovary.

Vascularization

The anatomical structure of the axis of the inflorescence and of the pedicel is similar to that of a typical stem; the vascular cylinder may be entire or split. Within the receptacle the shape of the stele becomes similar to that of the receptacle itself, i.e. the stele is usually widened at the base and narrows towards the upper part of the receptacle. From the stele of the receptacle the traces pass out to the various floral organs. The traces and the gaps split the stele into a characteristic network of vascular bundles. The stele may be further split as a result of the reduction of vascular tissues, as occurs in a typical stem. The traces of the various floral organs leave the stele in whorls or along a spiral line, in accordance with the arrangement of the organs on the receptacle. The traces of the uppermost carpels may, in certain plants, terminate the stele of the receptacle. In other plants, however, the stele continues above the level of the traces to the uppermost carpels. In this case the terminal bundles of the stele of the receptacle, which consist only of phloem or procambium, gradually fade out within the top of the receptacle.

The number of traces to the various floral organs varies. The number of traces to a sepal is usually equal to that supplying the foliage leaves of the same plant. Petals generally have one trace, but in some families three or more traces may be present. Stamens usually have a single trace but in some ranalian families and certain species of a few other families, e.g. the Lauraceae and Musaceae, three traces enter each stamen. Carpels may have one, three, five, or more traces; three traces is the most common condition, but five also occur frequently. The median carpel trace leaves the stele at a lower level than do the other traces. The continuation of this median trace constitutes the dorsal bundle of the carpel and the trace itself is termed the *dorsal trace.* The dorsal bundle is actually homologous to the leaf mid-rib.

The two outermost traces on each side are termed the *marginal* or *ventral traces,* as the bundles that pass along the carpel margins are continuous with them. If the carpels are fused to form a syncarpous gynoecium, these marginal bundles are found lateral to the dorsal bundle, and if the carpels are folded inwards they are found ventrally relative to the dorsal bundle. As a result of the inward-folding of the carpel margins the ventral bundles are inverted—their phloem is directed towards the lucules of the ovary, and their xylem outwards. If more than three traces are present in a carpel the additional traces are found between the ventral and dorsal traces, and they are termed *lateral traces.*

The vascular bundles of the carpels may branch and may even continue to do so during the development of the fruit. The vascular bundles that supply the ovules usually originate from the ventral carpel bundles, or from branches of them that are present in the placenta. The ovular bundle is single and thin, and reaches the zone of the chalaza; it does not penetrate into the nucellus but in some genera, branches of it enter into the integuments. The structure of the receptacular stele and the system of vascular bundles within the floral organs is, relatively complex even in the simpler types of flowers.

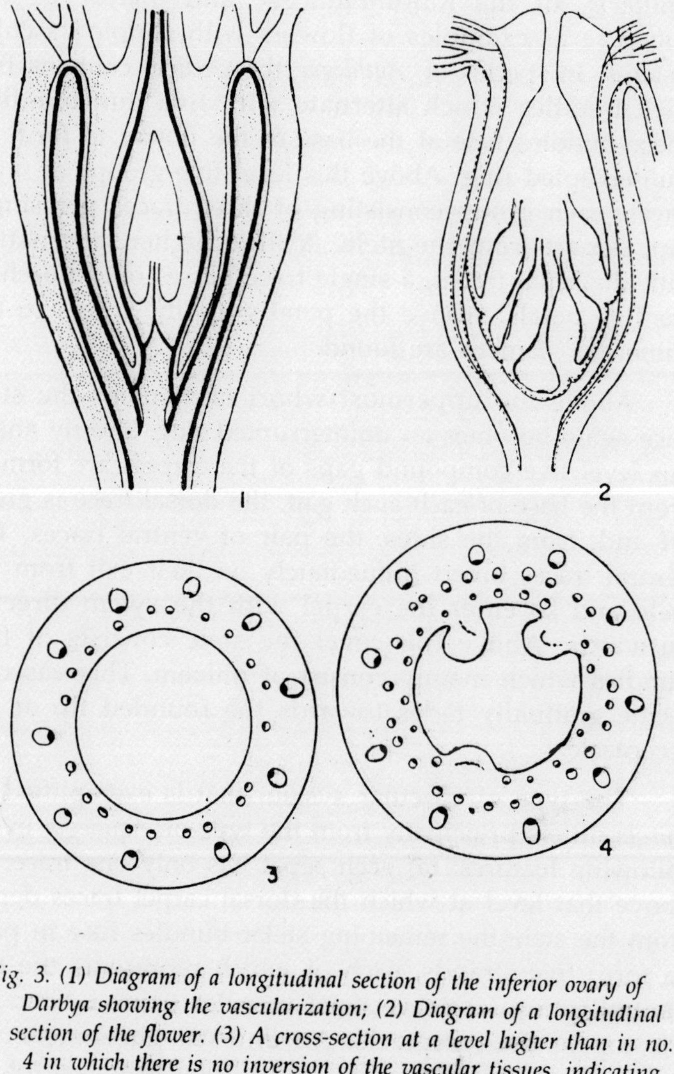

Fig. 3. (1) Diagram of a longitudinal section of the inferior ovary of Darbya showing the vascularization; (2) Diagram of a longitudinal section of the flower. (3) A cross-section at a level higher than in no. 4 in which there is no inversion of the vascular tissues, indicating that at this level there is no invagination.

The interpretation of floral vascularization becomes even more difficult in flowers in which fusion of the traces has taken place during the process of evolution. Eames gives *Aquilegia* of the Ranunculaceae and *Pyrola* of the Ericaceae as examples of flowers with simple vascular systems. In species of *Aquilegia* the pedicel contains five thick bundles which alternate with five thin bundles. These bundles fuse at the base of the flower to form an uninterrupted ring. Above this level five groups of sepal traces, each group consisting of three traces per single gap, depart from the stele. A little higher, alternating with the sepal traces, a single trace passes out to each of the five petals. Above the petal gaps the traces to the numerous stamens are found.

Above the uppermost whorl of stamens the stele once again becomes an uninterrupted ring. Shortly above this level five compound gaps of the carpels are formed. From the base of each such gap, the dorsal trace is given off and, from the sides, the pair of ventral traces. The ventral traces invert immediately on their exit from the stele and so enter the carpel with the xylem directed outwards. Above this level the stele consists of five bundles which mainly consist of phloem. This vascular tissue gradually fades towards the rounded tip of the receptacle.

The type of vascular system that is exemplified by *Pyrola* differs principally from the type of *Aquilegia* in the following features: (a) each sepal has only one trace; (b) above that level at which the dorsal carpel traces depart from the stele the remaining stelar bundles fuse in pairs to form five strands, each of which represents the two ventral traces; (c) no vestigial vascular tissue is found at the top of the receptacle. An unusual phenomenon of xylem discontinuity occurs in the tribe Psoraleae of the Leguminosae. The ovary bundles merge at the base of the ovary where they end in a horizontal proliferation of tracheary elements. This plate-like xylem expansion is not connected to the xylem of the pedicel. The xylem of the petal and stamen bundles in the receptacle is also commonly discontinuous. The phloem, however, is continuous throughout the flower.

During phylogenetic development of the flower, processes of cohesion, adnation and organ abortion have taken place. It is commonly accepted that during the process of evolution the external fusion of organs precedes that of the inner tissues, and that the fusion of the vascular tissue represents the last stage in this process. Fusion of vascular bundles involves those bundles that were close to one another. The fusion may involve the traces alone or may include part of the bundles within the organs. Infrequently it continues throughout the entire length of the bundles. Usually there is ontogenetic and histological evidence that fusion has taken place. Evidence of aborted organs is assumed from the presence of rudimentary organs and persisting traces as compared with related flowers.

As a result of the cohesion of organs, in many cases the lateral and marginal vascular bundles and traces fuse. In the calyces of plants of the Labiatae different stages of vein fusion can be found. Thus, according to Eames, in *Nepeta* and *Monarda* it is possible to discern that each sepal has a main vein and two lateral veins; in the calyces of *Ajuga* and *Physostegia* the neighbouring lateral veins are seen to be fused almost up to the base of the sinuses between the calyx lobes; *in Salvia* two pairs of lateral veins are fused while the others remain free. In sympetalous corollas vein fusion may also be found. In the sympetalous flowers of the Rubiaceae, for example, it is still possible to discern fifteen veins, three in each petal. In the Compositae it is possible to find evidence of the fusion of the lateral bundles and of the loss of the median bundles. In the gynoecium the fusion of the bundles takes place in the following ways. If the carpels are free the two ventral bundles approach one another on the ventral side of the primitive type of carpel that develops into a follicle. The ventral bundles may fuse from the base and the fusion may continue along the entire length of the carpel or along only part of it. A carpel with a single dorsal and a single ventral bundle may have two or three traces. In the former case one trace is related to the dorsal bundle and one to the ventral. In very reduced carpels, as, for instance, those in cypselae, there is sometimes only one trace, and it splits at the base of the locule of the carpel. In such a trace the three original traces are fused.

In *Filipendula,* for instance, which appears to be a highly evolved genus of the Spiraeoideae (Rosaceae), fusion of the carpellary bundles has occurred in such a way. If the carpels are fused, two types of fusion are distinguished. On the one hand, when a large portion of the carpel is folded inwards and the placenta is axile, the ventral bundles are brought to the centre of the ovary where a pair of inverted ventral bundles of a single carpel or those of two adjacent carpels fuse. The lateral bundles of two neighbouring carpels may also fuse in such an ovary.

On the other hand, if the carpels become fused while they are still open and so form a single common locule with parietal placentation, the ventral bundles are not inverted. They occur in pairs along those lines where the cohesion of the carpel margins takes place, or they may fuse to form common bundles. In addition to the fusion of bundles as a result of cohesion of organs, bundle fusion also results from adnation, i.e. the fusion of organs of different whorls. Bundles that are radially or tangentially close to one another may fuse, and this process may involve different numbers of bundles which may belong to two or more whorls.

The problem of the phylogenetic development of the inferior ovary still challenges flower morphologists. The problem is whether the gynoecium has sunk into a concave receptacle which then surrounded the carpels and fused with them *(axial theory)*, or whether the inferior ovary arose as a result of the cohesion and adnation of the lower portions of all the floral organs to the gynoecium, which they surround *(appendicular theory)*. In order to solve this problem different methods within the field of comparative morphology, involving ontogeny, histology, teratology and palaeontology have been used. Much attention has been paid to the floral vascular anatomy. The latter line of investigation in connection with the inferior ovary has revealed the following features. In the wall of the inferior ovary of certain plants, e.g. *Hedera*, separate traces which are related to the different floral organs are found.

In many plants, e.g., *Juglans* and a few species of the Ericaceae, different stages of bundle fusion have been observed in the wall of the inferior ovary. In relation to these plants, therefore, it is possible to conclude from the vascular anatomy that the wall of the inferior ovary consists of the ovary wall proper together with the other floral organs that are fused with it. In species belonging to the Cactaceae and San talaceae, inverted vascular bundles, i.e. those with inwardly directed phloem and outwardly directed xylem, can be found along the entire length of the inferior ovary wall.

This feature suggests that in these species the inferior ovary has developed as a result of the involution of the receptacle, as the sinking of the gynoecium into the receptacle necessitates an inward folding of the upper portion of the receptacular stele. In *Rosa,* Jackson found, according to the vascular anatomy, that the lower portion of the fleshy fruit consists of receptacular tissue, while the upper portion of it consists of fused floral organs.

Jackson also showed that various intermediate stages, which demon strate the evolution of the rose fruit, can be seen in related genera. The fruit of *Pyrus malus* var. *paradisiaca* and related genera apparently consist mainly of true ovary wall and of other floral organs that are fused to it, and the receptacle constitutes a very small portion, which is not easily discernible, at the base of the fruit. The adnation of the bundles in an inferior ovary that is developed from the fusion of the ovary wall proper to other floral organs is not equal in different radii, neither in the number of traces involved nor to the extent to which they are fused.

On the basis of floral vascular anatomy and floral ontogeny and histogenesis the inferior ovary of *Downingia* was also interpreted as wholly appendicular, representing adnation of the outer floral organs to the gynoecium. In addition to fusion between the floral organs of a single flower fusion may occur between flowers and the bracts accompanying them, e.g. *Lonicera* and *Juglans,* and even between several flowers, e.g. in certain species of *Lonicera, Cornus* and *Maclura.* These features are also often accompanied by the fusion of bundles.

Vestigial vascular bundles

The presence of vestigial vascular bundles, in certain positions, is frequently taken as proof of the degeneration of the organs that were originally present in the same position in the ancestral floral type. Vestigial petal traces may be found in the receptacle of apetalous flowers, e.g. *Aristolochia,* species of *Rhamnus, Salix* and certain species of *Quercus.* Vestigial staminal traces can be found in the Scrophulariaceae, Labiatae, Cucurbitaceae, and in certain genera of other families. In the Caprifoliaceae, Ericaceae, Rutaceae, and Valerianaceae vestigial carpel traces are found. Such vestigial traces may also be found in unisexual flowers. Degenerate ovules are commonly associated with vestigial bundles. The reduction that has taken place during the evolutionary specialization of the floral organs is often accompanied by the reduction of the vascular supply to the organs. This is exemplified by the development of the achenium or cypsela, which contains a single ovule, from a legume or follicle, which contain several ovules, and by the reduction of the corolla in the capitulum of the Compositae. Such reduction is recognizable by the shortening of the bundles. In some carpets, in which much reduction has taken place, only one bundle, which enters the ovule directly from the receptacular stele, is found. In stamens the bundles may sometimes terminate at the base of the filaments instead of reaching the anther. In the sympetalous corollas of many of the Compositae the median bundle is lost and lateral bundles may also become reduced, and they even disappear in the corollas of certain plants. The importance of the structure of the floral vascular system in the interpretation of the morphological structure and the clarification of the phylogenetic relationships between different plants is still debatable.

Histology of sepals and petals

In some plants sepals and petals were reported to initiate by periclinal divisions in the second and third layers and primordia of stamens in the third and fourth layers. Sepals and petals undergo apical, marginal, and intercalary growth, similar to foliage leaves. The external structure of sepals and petals may be leaf-like. However, the internal structure only of green sepals resembles that of foliage leaves and the internal structure of coloured sepals and petals is distinctly different. In the perianth the vascular system is usually only poorly developed and the veins usually lack sclerenchyma. The mesophyll commonly consists of spongy parenchyma only and its cells contain chromoplasts or pigments in the cell sap, or both. Usually the epidermal cell walls are thin. The anticlinal walls of these cells are, in many flowers, folded or undulated and are arranged so that the projections of the cells dove-tail one into the other, a feature that strengthens the epidermis. At the base of the petals, as well as along the length of the veins, the anticlinal walls are usually straight even when they are wavy in the other epidermal cells. The outer walls of the epidermal cells commonly have papillae, which make the petals glisten. More papillae are present on the adaxial than the abaxial epidermis and they are not developed at the base of the petals. Stomata, if present, are scanty and non-functioning. Trichomes may sometimes be present on sepals and petals. Sometimes intercellular spaces, covered by cuticle, are formed in the epidermis of petals.

Various pigments are found in the epidermal cells of the petals and sepals. The less specialized the petals are, the more their internal structure resembles that of foliage leaves, that is, the veins and the mesophyll are better developed, a palisade-like tissue is present, the epidermis is devoid of papillae and there are many stomata. Sepals and petals may fuse to form a lid or *operculum*. The sepals and petals may form two separate opercula or may be fused and form one common operculum.

Stamen

The stamens and carpels of highly specialized flowers differ greatly in structure and general shape fro m foliage leaves. Different opinions still exist as to the morphological interpretation and evolutionary development of the stamens and carpels. The most commonly accepted theory is that these organs are homologous to leaves. The opinion also exists that the development of the stamen can be explained on the basis of the *telome theory,* that is, from primitive dichotomously branched axes. According to this theory the stamen has developed as the result of the reduction and fusion of a system of axes that bore sporangia at their tips. New light was thrown on the phylogeny of stamens and carpels by Bailey and his co-workers from their morphological and comparative anatomical research in many families of the Ranales. From these investigations it has become apparent that in the woody species of the Ranales that exist today not only primitive stages of xylem development have been preserved, but also primitive types of stamens and carpels.

The epidermis of the filament possesses a cuticle and, in certain species, trichomes. The filament consists of parenchyma with well-developed vacuoles and small intercellular spaces. Often pigments are present in the cell sap. The size and external shape of angiosperm stamens vary greatly, but the anther generally contains four pollen sacs (microsporangia) which are paired in two lobes. The two lobes are separated by a zone of sterile tissue, termed the *connective,* through which a vascular bundle passes.

Complete and incomplete septa occur in the pollen sacs of some genera. According to Boke, who worked on *Vinca rosea,* even some time after the region of the developing anther becomes discernible in the staminal primordium, the anther still consists only of protoderm and a mass of ground meristem. All of the subepidermal layer of the young anther may be sporogenous, but actually the sporogenous tissue is developed from four cell regions which are located in the four angles of the developing anther. In each of these regions there is a row of hypodermal initials which divide periclinally to form two layers.

The inner layer of these initials constitutes the *primary sporogenous cells,* which by further division form the *pollen mother cells.* The outer layer of the above initials constitute the *primary parietal cells,* from which the wall of the pollen sacs and a large portion of the *tapetum* develop, as a result of periclinal and anticlinal cell division.

Seeds

In the seeds of certain plants, e.g. *Beta,* the nucellar tissue persists and increases in volume to form the *perisperm.* Several features can be distinguished on the outer surface of the seed. The micropyle may be completely obliterated or it may remain as a distinct pore. In the place where the seed was attached to the funiculus a scar, termed the *hilum,* is present. Water can penetrate with relative ease through the hilum. In anatropous ovules, in which part of the funiculus is fused, to the integument, the seed becomes detached together with the fused part of the funiculus which forms a characteristic ridge termed the *raphe.*

After the fertilization, of the ovule, growths termed *arils,* develop on the surface of the seeds of certain plants. Arils are very common in tropical and subtropical plants. e.g. in *Intsia bijuga, Pithecellobium dulce, Durio zibethinus, Acacia retivenea, Bocconia frutescens,* and *Myristica fragrans.* Arils are organs that are well adapted to seed dispersal by animals. Arils that contain oil, such as those of Chelidonium *majus, Luzula villosa, Gagea lutea, Reseda odorata, Galanthus nivalis, Dendromecon,* and *Ricinus communis,* are elaiosomes. They are thought to be connected with seed dispersal by ants.

For structures adapted for seed dispersal see Fahn and Werker. Some authors have distinguished between true *arils* and *arilloids.* True arils are outgrowths of the funiculus near its top, whereas arilloids are outgrowths from other parts of the seed. Arilloids occurring on the raphe are called *strophioles* and those occurring near the micropyle *caruncles.* Another type of seed appendage is the plug-like structure, formed in the micropylar region, that separates during germination by circumscissile dehiscence. This type of appendage is termed *operculum.*

Angiosperm ovules have one or two integuments. In order to clarify which parts of the integuments take part in the formation of the seed coat, ontogenetic investigation is necessary. All parts of one or both integuments may take part in the formation of the seed coat. e.g. *Viola tricolor*. However, in most seeds much of the integumental tissue is destroyed and absorbed by the other developing tissues of the seed, and then the seed coat develops only from the remaining parts of the integuments. The parts that are destroyed are usually the innermost or intermediate middle layers of the integument. The nucellus may also take part in the construction of the seed coat.

However, in the development of most seeds the nucellus is apparently completely destroyed. In certain seeds, especially those of indehiscent fruits, only the two or three outermost layers of the integument persist. In the mature seed of the Umbelliferae only the outer unspecialized epidermis of the single integument remains. In the seeds of certain genera of the Compositae, e.g. *Lactuca*, the integument is represented only by a thin layer of obliterated cells which persists beneath the cypsela coat. The innermost layers of the latter coat also disintegrate.

In the seeds of certain monocotyledons, e.g. *Zea*, the integuments are completely destroyed. In seeds developing from ovules with two integuments, the two integuments may be present in the testa or only two or three of the outermost layers of the outer integument may remain. In such seeds the inner integument may constitute the major portion of the seed coat, or the outer one may be the better developed and be specially adapted for protection while the inner is non-specialized. The former type is characteristic of the Malvaceac, Violaceae, Hypericaceae and Tiliaceae, and the latter of the Cruciferae, Papaveraceae, Berberidaceae and certain genera of the Liliaceae, Iridaceae, and Araceae. In the Onagraceae, Lythraceae, Aristolochiaceae, and others, protective layers develop from both integuments, and the entire nucellus or its outermost layers only also take part in the formation of the seed coat. In most of the Leguminosae, Ranunculaceae and certain genera of the Liliaceae and Amaryllidaceae the inner integument and the nucellus are completely obliterated. In the seeds of only a few plants in which the ovule is surrounded by a single integument does the entire integument form the seed coat.

Mucilage is common in many seeds, nutlets, and one- seeded fruits. It appears in Cruciferae, Labiatae, Compositae, Plantaginaceae, and other families. The functions of mucilage are many: when the diaspore is moistened after its release, it may adhere to the soil surface this feature has the advantage that the adhering diaspore cannot be carried farther away to unfavourable localities by wind or rain; the diaspore may adhere to animals and be thus further dispersed; the mucilage may cause reduction of the specific weight of the diaspore in water; the mucilage may take part in regulation of germination by preventing desiccation of the germinating seed or, in case of excess of water, when the seed becomes entirely covered by mucilage, the mucilage may hinder passage of oxygen, thus preventing germination.

Endosperm tissue

Endosperm tissue in the developing seed may consist of thin-walled cells with large vacuoles which do not contain reserve substances. Such an endosperm is entirely or partially absorbed by the developing embryo. e.g.

Lactuca. In many other seeds, e.g. those of *Ricinus* and the Gramineae, the endosperm functions as a storage tissue. An endosperm of this type may consist of thin-walled cells and then the reserve material is located within the cell, or it may consist of thick-walled cells and then the walls themselves constitute the reserve substance.

There are two main forms in which proteins may be stored in the endosperm—in an amorphous form (glutens), or in the form of aleurone grains. Aleurone grains are composed of an amorphous protein mass. Some enclose crystalloids and some spherical bodies which may contain phytin, others contain both crystalloids and globoids while still others contain none of them. In the caryopses of cereals glutens are found in the starch cells, and aleurone grains are restricted to the outermost layer of endosperm cells.

Fruit

Fruits of many plants, such as certain varieties of *Musa, Citrus,* and *Vitis,* develop without the formation of seeds. This phenomenon is termed *parthenocarpy.* In some plants, e.g. in the peanut, the fruit develops only after the gynophore penetrates into the soil carrying the carpel with the fertilized ovules. Some plants produce fruits of more than one kind on the same individual. This phenomenon is called *heterocarpy. Aethionema heterocarpum* and *Ae. carneum* (Cruciferae), for instance, produce dehiscent siliquas and one-seeded indehiscent nuts in the same raceme. Cases in which different fruits are borne above and below the ground are termed *amphicarpy,* e.g. *Trifolium polymorphum, Vicia angustifolia* var. *amphicarpa, Gymnarrhena micranta* (Compositae), and *Emex spinosa* (Polygonaceae).

Pericarp

During the development of the fruit the number of cells increases and the parenchymatous ground tissue may remain as it is, or part of it may become sclerenchymatous. In the mature fruit the above three layers—the exocarp, mesocarp, and endocarp—can be distinguished but sometimes only the exo- and endocarps are discernible. However, in many fruits the exo- and endocarps may consist only of epidermal tissues and the pericarp then consists mainly of mesocarp. The division into these pericarp layers is only of convenience to facilitate the anatomical description of mature fruits, and these layers do not represent separate tissues from the point of view of their origin.

Chapter 8

Phototropism in Plants

The literature of phototropism contains a diversity of experimental organisms and experimental conditions. While the former is as inescapable and the diversity of life itself, the latter is more subject to control. In the hope that it will provide useful guidance to the reader, a brief discussion of some of this experimental diversity is offered. The response consists of the change in orientation of the phototropic organ with time. Ideally, the data should consist of the recording with time of the shoot axis angle in three-dimensional space. Interestingly, Darwin did take the trouble to record the movement of shoots and coleoptiles as seen from above.

Rather few studies since then have followed this example, perhaps because of the complex repertoire of nutational movements Darwin discovered. Most phototropic data are derived from angles projected onto a plane, under the reasonable but usually unproven assumption that little significant information is thereby lost. How long should this bending be followed? In oat or corn coleoptiles starting from a vertical position, interference' by the geotropic response becomes apparent after 100 min or so. One simple but arbitrary approach is therefore to record the angle at 100 min. The difficulty is that the method will not distinguish the speed of bending from the maximum accumulated angle of curvature since phototropic organs frequently differ in both of these measures, it is important to use consistent methods in making comparisons. Another attractive but experimentally more difficult approach is to use a clinostat to eliminate geotropism altogether. Using this method, both the rate of bending and the total curvature in response to a discrete stimulus may be measured with considerable confidence.

Blaauw was one of the first to abandon the more natural conditions used by Darwin and investigate the effect of short light flashes. Blaauw was interested in the validity of the Bunsen-Roscoe reciprocity law, and in fact he found that to obtain a threshold response from *Avena coleoptiles*, a constant energy fluence (= fluence rate X time) was required, and that this constancy held over a range of exposure times from 43 h to 1 ms. This simplified view no longer prevails, however, and long exposure times are now known to trigger processes that lea For example, the classical third positive curvature does not obey the reciprocity law and is probably due to the effect of blue light in altering the sensitivity of the system. This 'tonic' effect evidently is time limited instead of energy limited.

There are many variations in the handling of experimental material prior to the beginning of the experiment, although in the majority of cases dark-grown seedlings, handled in 'dim' green light, are used. One of the most persistent problems is that the effect of a red light pretreatment is overlooked. Not only is it risky to compare results from dark-grown seedlings with those from seedlings exposed to a red pretreatment, it is also questionable to assume that 'white' or 'blue' phototropic simulus sources have too little contaminating red light to affect the results.

Typical fluence (dose)-response curves were obtained by Zimmerman and Briggs using initially vertical *Avena* coleoptiles pretreated with 2 h of red light. They exposed the apical 3 mm to three fluence rates (intensities) of monochromatic blue light at 436 nm, and at each fluence rate the total fluence was varied by varying the exposure time from 0.7s to 37 min. The curvatures were measured at 100 min after the end of the exposure. The curve at the highest fluence rate shows the characteristic features noted by earlier workers. The initial rising part of the curve and the first peak are termed the first-positive curvature, the subsequent minimum is termed the first-negative curvature, and the second rising part is termed the second-positive curvature.

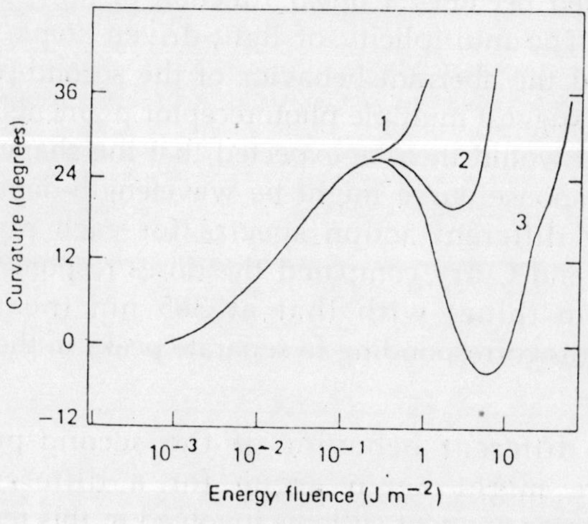

Fig. 1. Fluence—response curve for phototropism in Avena coleoptiles.

As the fluence rate is reduced, the region of the second-positive curvature shifts towards lower fluences and in the process the first-negative region becomes obliterated. For each 10-fold reduction in fluence rate, the fluence required to produce a given second-positive curvature, say 36°, is likewise reduced 10-fold. In other words, this curvature is characteristic of a certain exposure time rather than a particular fluence, and hence reciprocity does not hold.

Zimmerman and Briggs proposed an empirical kinetic model in which first-positive and first-negative curvatures are each explained by simple two-step reactions with a pair of photosensitive forward rate constants. By appropriate adjustments of the four arbitrary constants the authors explained their experimental results in the region of the first-positive and first-negative curvature.

In the region of the second-positive curvature, they proposed a very complex scheme, comprising a photoequilibrium step followed by a thermochemical step, so that at suitably long times the amount of product accumulated becomes a linear function of time, but not intensity. The multiplicity of light-driven steps in this model and the aberrant behavior of the second-positive curvature suggest multiple photoreceptor pigments. If this were so, it would then be expected that the shape of the fluence-response curve might be wavelength-dependent, reflecting different action spectra for each pigment. Thimann and Curry compared the dose- response curve at 436 nm (blue) with that at 365 nm (near UV), wavelengths corresponding to separate peaks in the action spectrum.

The different behavior of the second-positive curvature might again argue for a difference in photoreceptor pigment systems involved in this response. However, the action spectrum for the second-positive curvature in *Avena* is not different from that for the first-positive curvature and it must be concluded that all regions of the fluence-response curve derive from excitation of a single photoreceptor. Although the complexities of the kinetic model of Zimmerman and Briggs do not correspond to complexities in the primary photoreceptor system of the phototropic response, this of course in no way invalidates the model, since its features could well be the result of many other kinds of processes, removed from the primary photo-event.

One of the difficulties in understanding the fluence-response relations of coleoptiles is that the measured curvature is a function of two variables: the fluence rate and the duration of the exposure. The fluence-response relationship must therefore be visualized as a three-dimensional surface, rather than a two-dimensional curve. Blaauw and Blaauw-Jansen have measured curvature of A vena coleoptiles as a function of fluence rate from 10^{-5} to 12.6 W m^{-2} and as a function of exposure time from 0.01s to 40 min, The coleoptile was pretreated with red light, and the curvature was measured 100 min after the end of the exposure. The data have the usual appearance when plotted either as a function of fluence rate or as a function of time, holding the other factor constant in each case. However, when the curvature is displayed as a function of two variables at once, a three-dimensional surface results. Although it has a rather unconventional appearance, this surface may be the most meaningful way to comprehend the fluence-response relation as a whole.

These sections yield response as a function of time and differ markedly when made at different fluence rates. At the highest fluence rates (12.6 W m^{-2}) the curve is quite complex, showing three maxima known classically as first-positive, second-positive, and third-positive curvature, and designated by Blaauw and Blaauw-Jansen as areas A, B, and C, respectively. As sections are made at progressively lower fluence rate values, areas A and B shift to longer exposure times, as does the negative region between A and B (the classical first-negative response). This shift is in accordance with the reciprocity law; for example, the peak curvature of area A corresponds to a constant energy fluence (given by the authors as 0. 12 J m^{-2}) equal to the product of time and fluence rate. Area C does not shift along the time axis; rather, its peak and slope increase with decreasing fluence rates, showing that reciprocity does not hold.

This pattern of changing curve shapes has the intriguing result that at progressively lower fluence rates area B becomes merged with, and finally obliterated by, the rising flank of area C. The negative region, too, becomes swallowed up and finally vanishes at a fluence rate of about 10^{-3} W m^{-2}. At shorter exposure times the sections do in fact correspond to the classical first-positive, second-positive curvatures, areas A and B respectively. For example, imagine the surface of sectioned by a vertical plane at an exposure time of 15 s. The fluence-response surface relationship is a convenient way to synthesize a great variety of experimental results into a single pattern. It also allows the disentangling of the second-positive and third- positive responses, and it indicates what conditions of energy fluence rate and exposure time are best suited for their individual study.

The complexity of the fluence- response relationship implies that each energy increment of the phototropic stimulus has two effects. It triggers a bending response, and it also changes the state of the organ so that its sensitivity to a subsequent energy increment is now altered. The latter is a tonic, or non-directional effect of blue light. In *Avena*, the first stimulus can be an effective desensitizer even when given vertically from above. Blaauw and Blaauw-Jansen found that the C-type (third - positive) curvature obtained with a 10-min exposure to blue light (fluence rate 1.48×10^{-2} W m^{-2}) could also be elicited if as much as the first 9 min of the exposure was given vertically instead of horizontally.

The vertical exposure could be shortened to 1 min, provided that a dark period of at least 2 min intervened before the subsequent horizontal exposure. This dark period could be as long as 40 min without diminishing the positive (C-type) curvature in response to the second exposure. Since the horizontal 1-min exposure taken by itself is in the first-negative region, the increased curvature can be interpreted as a reactivation of the first-positive response caused by a lowered sensitivity. Thus the C-type curvature is not actually an independent response at all, but the result of desensitization of the first-positive response system. This desensitization requires from 5 to 10 min to become fully developed after which it gradually dies out. This process can be visualized as the movement of the A-region, or first-positive curvature, towards higher fluence rates, reaching a maximum shift 5 to 10 min after the first exposure, and then moving gradually back to its original position.

The first-positive region in *Avena* was shifted to higher fluence rates by red light. These results were confirmed by Zimmerman and Briggs, who showed that the regions of first-positive and first-negative curvature were shifted to higher fluence rates (10-fold shift) by a red-light exposure immediately prior to blue-light exposure. They found that the second-positive region was shifted towards lower fluence rates (3-fold shift) by this same red-light pretreatment.

Thus the red pretreatment lowers the sensitivity of the first-positive and first-negative response, and raises the sensitivity of the second-positive response. In terms of the fluence -response surface, red pretreatments shift the A-area curvature toward higher fluences and the C-area curvature toward lower fluences which amounts to an increase in C-area curvature. A different situation exists for area B, which disappears completely if no red pretreatment is used, leaving only a broad indifferent region between the two minima.

In addition to an increase in the second-positive (C-type) curvature, red light enhances the rate of elongation near the coleoptile tip. This change in elongation distribution may be related to the effect of red light on the localization of curvature. When blue stimuli in the fluence range of the C-type curvature were used, red pretreatment shifted the locus of maximum curvature towards the coleoptile tip. This distribution of red-light sensitivity along the coleoptile is fairly well correlated with the distribution of phytochrome: both were most heavily concentrated in the apical 3 rum.

Fluence- response comparison of the two red-light effects had quite unexpected results, however. The sensitivity of the physiological effect on phototropism is a factor of 10^3 or 10^4 higher than that of the effect on spectro-photometrically detectable phytochrome. This enormous difference in sensitivity indicates that if both effects are indeed due to phytochrome, the pigment must be divided into two pools: a large spectrophotometrically detectable one, and a small spectrophoto metrically undetectable one that is, however, physiologically active.

Further data collected by Briggs and Chon, concerning the decay kinetics of P_{fr} and P_f and the shape of the fluence -response curves, indicate that if the same phytochrome pigment is responsible for both effects, the pools must be functionally distinct, and the photochemical rate constants for P_r-P_{fr} interconversion must be different within each pool.

Perception of stimulus

Analysis of a photobiological process requires the identification of the primary photoreceptor pigment, and the first step in such identification is the determination of an action spectrum. The action spectra for the blue-light type of response (in which red is phototropically ineffective), typical of coleoptiles, Phycomyces sporangiophores and many other systems, will be described first. There are two prominent peaks and a shoulder in the blue region (475, 450, and 420 nm), and there is a peak in the near-ultraviolet region (370 nm). The A vena action spectrum in the ultraviolet region shows a prominent peak at about 290 nm *Phycornyces* action spectra are determined most advantageously with a balance method, since the adaptive mechanism makes it impossible to obtain the usual fluence-response curve.

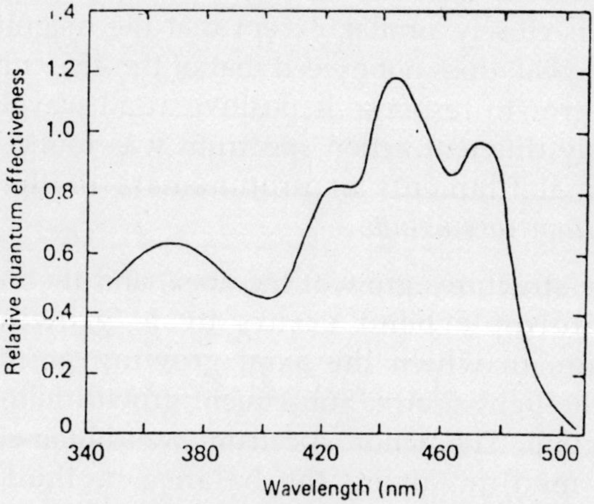

Fig. 2. Action spectrum for phototropism in Avena coleoptiles, showing the quantum effectiveness for a curvature of 10°.

One method balances a standard stimulus against the monochromatic stimulus, each source being located at one end of a 2-m plank. Cultures are placed along the plank, and the point of balance is determined. Action spectra for the visible, in which the standard stimulus is a source with peak wavelength at 460 nm, and for the ultraviolet *Phycomyces* sporangiophores show negative curvatures at ultraviolet wavelengths.

This reversal of Phycomyces phototropism is due to the strong ultraviolet absorption by the sporangiophore. There is an indifferent region, where neither positive nor negative curvatures occur, between 310 nm and 334 nm. If a single standard is used in both the ultraviolet and visible ranges, the resulting curves can be quantitatively compared. Using this method, Delbrück and Shropshire found that the UV peak at 280 nm is about four-fold higher than the main peak at 455 nm.

Here also, the reason ties in the strong absorption by the cell at 280 nm, leading to an exaggerated intensity gradient across the cell. The growth-response action spectrum is closely similar, except that the magnitude of its 280-nm peak does not exceed that of the 455 - nm peak. The light-growth response is positive at all wavelengths. A strikingly different action spectrum was found for the chloronemal filaments of protonemata of the moss, *Physcomitrium turbinatum.*

These structures grow at the apex, and the initiation of a phototropic bend occurs by a 'growing out' mechanism, in which the axial growing point shifts towards the light source. Subsequent growth follows this new direction. The action spectrum was obtained in an aqueous medium using the balance method and a standard source of wavelength 665 nm. Blue light is almost totally without phototropic effect, although there is a slight suggestion of activity at 365 nm. Phototropism in moss protonemata thus uses a different photoreceptor pigment than in Phycomyces and most higher plants.

Photoreceptor pigment

Avena coleoptiles and *Phycomyces sporangiophores* have very nearly identical action spectra, and this strongly indicates that both organisms have the same phototropic photoreceptor pigment. Many other action spectra are similar and are probably due to the same blue-absorbing pigment. This similarity suggests that there may be a unity in phototropic mechanism for these systems, and that such unity is built upon a unity of photoreceptor pigment. Is this pigment a flavin or a carotenoid? There is little basis for choice in the visible region, where the absorption spectra of both flavins and carotenoids are generally a good fit with the action spectra.

Carotenoids have a characteristic triple-peak structure in the region around 450 nm, and although riboflavin absorption spectra shown in earlier reports are devoid of fine structure in this region, spectra obtained in ethanol at 77 K do indeed have the requisite shape. More pronounced discrepancies show up in the near ultraviolet, however, where the riboflavin absorption peak in ethanol at 77 K is shifted toward shorter wavelengths by about 30 nm, as compared to the action spectrum. The carotenoid absorption spectrum typically lacks peaks altogether in the 300-400nm range. Both pigments generally have another peak near 280nm, although the magnitudes differ. Thus neither flavins nor carotenoids can be ruled out conclusively on the basis of these spectral comparisons.

Action spectra have been obtained for the Phycomyces light-growth response, the transient change of growth rate as a result of a change in the steady-state light intensity. Using a null point method, Delbrück and Shropshire obtained a growth-response action spectrum essentially identical to the phototropic spectrum, except that the size of the 280 nm peak was different, due to the screening effect of UV-absorbing substances. This work has been extended by Delbrück, Katzir and Presti in the region from 57-5 to 630nm, using a tunable laser.

Meissner and Delbrück analyzed the carotenoids of an albino *Phycomyces* strain with normal phototropism and were unable to detect any β-carotene. Using methods 10 times more sensitive, Presti, Hsu and Delbrück have detected no β-carotene in albino *Phycomyces* mutants doubly blocked for β-carotene synthesis but with normal phototropism. Thus, the photoreceptor in Phycomyces is probably not β-carotene but a flavin.

Moss protonemata are phototropically responsive to red light, suggesting the involvement of a red-absorbing pigment such as phytochrome, rather than the more common blue-absorbing photoreceptor pigment. It has recently been discovered that *Phyconzyces* shows phototropism to red light, with a very weak but unmistakable maximum in the action spectrum at 650nm. Although this red peak is similar to that found in the growth action spectrum, comparison of the phototropic action spectra of wild type and,night blind' mutants suggests that the red phototropic peak is due to a second photoreceptor pigment and not to the excitation of a flavin triplet state in the primary pigment. An interesting red-blue interaction has also been described in Phycomyces by Löser and Schäfer.

When red light (605 nm) is combined with blue light (450 nm), the phototropic response is less than that for blue light alone. Thus it appears that red light has a dual effect; it can either trigger phototropism when acting alone, or it can inhibit the phototropic response to blue light. Although this could be the result of a single photochromic pigment like phytochrome, the mutant evidence favors two independent pigments. It is not known whether such red -sensitivity may exist in coleoptile phototropism. Since the size of the red peak in the Phyconzyces action spectrum is about 10^8-fold smaller than the main blue peak, a comparable red peak may well have gone undetected in previous work with coleoptiles.

Orientation of photoreceptor

Locus of maximum phototropic sensitivity of coleoptiles is thought to be the apex. This sensitivity is not confined to the extreme tip, however, but extends downwards for a considerable distance. For example, Meyer obtained phototropic responses by stimulating either the apical 350 µm of the coleoptile or the next 350 µm-segment with 5-

s blue light stimuli. Although curvatures were twice as large when the apical segment was stimulated, the fluence -response curve was not shifted along the fluence axis, but was merely changed in amplitude. This suggests that the sensitivity of the more basal section cannot be attributed to scattered light reaching the apical section indirectly, but must instead be due to the existence of a pigment in the more basal section.

If continuous illumination is used instead of short flashes, significant phototropic curvatures can be obtained even if as much as 5 mm of the apex are covered by an opaque cap. Although Curry observed only 'small' curvatures in capped coleoptiles after 100 min, much greater curvatures can be obtained after several hours, especially if geotropism is suppressed by the use of a clinostat. The use of microbeam stimuli has shown that the phototropic sensitivity of Phycomyces sporangiophores extends from 0.1 to about 2 mm below the sporangium.

Page and Curry found that the phototropic sensitivity of young sporangiophores of the fungus *Pilobolus kleinii* was confined to the terminal 50 μm of the cell. Illumination of the more basal region, which has a high concentration of a carotenoid pigment, produced little or no phototropic curvature. Thus in these cells the photoreceptor is much more localized than in Phycomyces.

An extremely localized receptor is also found in the alga *Vaucheria geminata*, which grows only at the extreme tip and responds to a phototropic stimulus by a lateral shift of the growth center. Kataoka gradually increased the fraction of the hemispherical tip illuminated by the half field and found that bending reached only 50% of normal when all but the terminal 20% was illuminated. This shows that most of the photosensitivity is concentrated in the extreme tip of this cell, perhaps the apical 10 µi. The existence of ordered, dichroic photoreceptors in Phycomyces sporangiophores has been well established.

Under conditions of high optical density, Jesaitis found that transversely polarized light was more effective than longitudinally polarized light in eliciting the growth response. He further concluded that the orientation of the blue and ultraviolet dipole moments of the receptor molecule are 5.5° and 33.5°, respectively, from the transverse (hoop) axis.

Dichroic, oriented photoreceptors have also been postulated in Phycomyces sporangiophores by Hertel and by Wulff but their measurements of the effect of the plane of polarization on the time of onset of phototropism seem to support a radial, rather than a transverse orientation. In moss protonemata, the finding that the phototropic effectiveness of red or far- red light is maximal when the plane of polarization of the 45° incident beam is vertical, leads to the conclusion that the receptors are near the cell surface and the preferred dipole orientation is parallel to it.

Signal processing after perception

The phototropic response is directed toward or away from a source of light, and this response must be based on the distribution of light intensity set up within that organ by purely optical mechanisms. Such a distribution will not be a simple gradient of intensity across the organ if it is created principally through optical refraction by the cell. On the other hand, transmission losses due to internal scattering or absorption will lead to a simple gradient.

The lens property of the sporangiophore of Phycomyces has been shown to be the principal means of establishing the intensity distribution. This was proved by the reversal of the direction of phototropism when the cell is immersed in a medium of higher refractive index but Shropshire demonstrated it still more elegantly by showing that the illumination of a sporangiophore with a diverging ray bundle also reverses phototropism. The receptors are apparently able to detect this intensity distribution by virtue of their circulation around the periphery of the cell in synchrony with the cell wall's spiral growth.

The introduction of an artificial rotation in the opposite direction greatly reduces the strength of the phototropic response. Phototropism in A vena coleoptiles is also reversed by immersion in paraffin oil, but it is not clear whether the mechanism is that of a focusing effect. Meyer found that soaking in oil prior to unilateral illumination in air also produces rev al. She found that reversal occurs only when this oil treatment is given at least 1 min before the 5 - s light stimulus.

Penetration of the oil into the intercellular spaces and into the upper 300 μm of the central cavity was observed microscopically. Although Meyer concluded that the reversal was due to physical changes in the cell wall structure, it also seems possible that optical effects may play a role at the cellular level in coleoptiles.

Direct evidence for a focusing mechanism in *Avena* was provided by Shropshire, who reported that coleoptiles showed negative phototropism when exposed in air to a diverging beam of light, using a fluence in the region of first-positive curvature. A converging beam at the same fluence produced positive phototropism.

In the absence of focusing effects, the light gradient across the cell will be determined by light absorption and scattering due to the cell contents. Although in *Phycomyces* the focusing effect is dominant, there are experimental conditions under which the screening becomes large and phototropism reverses sign.

The reversal in ultraviolet light is due to the high concentration of gallic acid in the cell which is strongly absorbant at these wavelengths. In addition, a mutant strain with a superabundant amount of b-carotene has an optical density about ten times normal and shows negative phototropism in blue light. Screening is important in moss protonemata, in which the cell apex contains high concentrations of oniorophyll. Apparently the lens effect is of no significance in this case, and phototropism is caused by a 'growing out' phenomenon on the illuminated side of the filament apex. In support of this are the observations that light stimulates the growth rate and that there is no reversal of phototropism in paraffin oil.

Following the electronic excitation of the photoreceptor molecule, but before changes occur in the distribution of auxin, there must intervene a sequence of signal - processing events. A clue to these events may be found in studies of light-induced biopotential changes, although there is no easy way to distinguish signals that relate directly to the causal sequence from those that are merely byproducts.

Coleoptiles develop transverse electric potentials when illuminated unilaterally, but the phenomenon is probably not related in any way to signal processing. Johnsson found that with a unilateral stimulus of up to 10 min the potential across a corn coleoptile does not appear until 30 min after the stimulus, suggesting that the potential may be a result of the auxin redistribution rather than its cause. A light-induced signal of short latency in *Avena* coleoptiles has been briefly reported by Briggs.

The potential between the tip and the base of the coleoptile changed within the beginning of a blue-light stimulus and could well be related to early intracellular signal processing. Interesting light-induced biopotential changes have been observed by Hartmann and Schmid in etiolated bean hooks. Biopotential changes occurred within 1-5 s after the beginning of the light pulse and reached a maximum within 2 min. Although not phototropism, hook opening is nevertheless a light-induced growth phenomenon, and it has a typical blue - light action spectrum.

Effects on metabolism

There have been several reports of light-induced changes in ATP and cyclic AMP (cAMP) levels in *Phycomyces* sporangiophores. But these findings have been difficult to confirm. Cyclic AMP has also been implicated in Vaucheria phototropism by Kataoka, who found that it greatly reduced phototropic bending without affecting the linear growth rate. Kataoka proposed that the cAMP functions in the cell's.steering mechanism', which can be swamped by an excess of the substance. The use of low temperature is an attractive approach to the study of signal processing because in this way it may be possible to separate the primary photo-event from the subsequent processes.

In *Avena* coleoptiles, a 45-min unilateral blue-light stimulus delivered at 2°C was found to remain in 'cold storage' for at least 450 min of additional time in the cold. Data of these authors also suggest that some signal processing occurs even at 2°C: at 25°C a 10-min exposure produced a 19° curvature, whereas at 2°C a 30-min exposure at the same fluence rate was required for the same curvature. Thus at least one step subsequent to the primary photochemical one continues in the cold, but at a lower rate. Other steps must be blocked almost completely at 2°C.

Briggs used the standard *Avena* curvature test to determine the auxin content of agar blocks placed beneath the basal end of illuminated corn coleoptile tips. In the range of either the first-positive or the second-positive curvatures, there was no significant difference between the auxin yields of the illuminated tips and those of the dark controls, thus indicating that under the usual conditions for phototropism no significant photo-destruction of the total diffusible auxin occurs. When coleoptile tips were used that had been sectioned longitudinally from the base up to, but not including, the extreme apex, and into which thin class barriers had been inserted between the sectioned portions, it was found that the shaded half of the tip delivered significantly more auxin into the receiver block. If the tips were completely divided and the halves totally separated by a thin glass barrier, no auxin differences were noted, thus indicating that some substance must move laterally in the intact part of the apex.

Lateral asymmetry of indoleacetic acid (IAA) has been demonstrated by Pickard and Thimarm, who applied ^{14}C-IAA to the extreme apex of excised corn coleoptiles in which the basal 1 mm was partitioned by razor blades. When a unilateral white light fluence corresponding to the first-positive curvature was administered, the radioactivity recovered in the basal agar blocks was in the ratio of 25:75, for the illuminated and shaded sides, respectively. Under similar conditions the radioactivity of the illuminated and dark halves of the coleoptile tissue itself had the ratio of 35:65. Further, there was no evidence for any effect of illumination on the total radioactivity recovered, ruling out a photo-inactivation mechanism. Thus, in this system at least, the IAA applied to the apex becomes asymmetrically redistributed, establishing a 'concentration' gradient in the tissue itself, and finally leading, via basipetal transport, to a concentration difference in the agar blocks.

Since in these experiments the unilateral light stimulus was administered before the ^{14}C-IAA was applied, the light clearly does not act on the applied IAA itself, but rather sets up in the tissue a lateral polarity, which subsequently leads to an asymmetric distribution of IAA. Several workers have noted rather variable results when the IAA application was not rigorously limited to the extreme coleoptile tip, suggesting that the site of origin of the asymmetric redistribution is highly localized at the apex. Another approach to this problem is that of a clearly off center IAA application, which gives an unequivocal test of lateral transport. Using this method, Gardner et al. found good evidence for lateral transport: intact corn coleoptiles exposed to blue light producing 11.60 curvature in controls had 31.1% of the radioactivity on the side opposite the point of application when the point of application was illuminated and 15.8% on the opposite side when the point of application was in darkness. These results indicate a 'net lateral movement' of 15.3% from the illuminated side to the dark side.

Unfortunately, this technique did not always give consistent quantitative correlations between radio activity differentials and curvatures, notably when comparing intact and excised tips and when introducing a red pretreatment. This may be due to a problem with the technique, or to the existence of hormones other than IAA capable of causing differential growth. In excised third internodes of pea seedlings (*Pisum sativum*), Kang and Burg were able to show that ^{3}H-IAA applied to the apical end of the section in lanolin paste became laterally asymmetrically distributed in the ratio of 35:65 (light: dark) under conditions where a phototropic curvature of 24° developed.

Remarkably, a red pretreatment that nearly doubled the curvature did not significantly change the lateral IAA gradient, again show curvature does not correlate quantitative with IAA gradients. In light-grown sunflower seedlings (*Helianthus annuus* L.), Bruinsma, Karssen, Benschop and Van Dort found no detectable gradient of IAA between the concave and convex flanks of phototropically bending plants when the flanks were extracted and analyzed for IAA by a spectrophotometric method.

The cut basal end of the hypocotyl secreted considerable auxin into a water or buffer solution, as measured by *Avena*-test activity. However, when spectroscopic IAA determinations were made on these diffusates, only a few per cent of this auxin activity could be accounted for. Thus, IAA is probably not responsible for differential growth in Helianthus. In the same organism, Phillips found a ratio of 12:88 (light: dark) in the diffusible gibberellins extracted from unilaterally illuminated excised hypocotyls. Hormone activity was measured on chromatographed extracts using a lettuce hypocotyl elongation test.

Chapter 9

Researches in Photosynthesis

Research into photosynthesis began more than 200 years ago, when Priestley discovered that photosynthesis was the reverse of respiration. The discoveries of the next century indicated that water, carbon dioxide, light and the green parts of plants were all necessary for photosynthesis. The green pigment chlorophyll was isolated early in the nineteenth century, and later the hypothesis that photosynthesis led to the storage of chemical energy was proposed. There followed careful work on the stoichiometry of gas exchange, and the finding that starch was the product of photosynthesis.

All higher plants and algae, together with certain species of bacteria, are capable of storing light energy via photosynthesis. The most primitive photosynthesizing cells are the photosynthetic bacteria, and their photosynthesis differs from that of other organisms because they are not able to evolve oxygen and because they contain the pigment bacteriochlorophyll, rather than chlorophyll as in green plants. They are, however, widely used in photosynthesis research because the cells are amenable to a range of biochemical and molecular biological treatments. Mainly for this reason, our knowledge of the photosynthetic processes in bacteria is more advanced than that of higher plants. The link between bacterial photosynthesis and green–plant photosynthesis is provided by the blue–green algae . These organisms have the characteristics of prokaryotic cells, but have the ability to evolve oxygen; their pigments are also more like those of green plants, since they contain chlorophyll.

Oxidation and reduction

van Niel studied the comparative biochemistry of photosynthesis in oxygen–evolving organisms and in the non–oxygen-evolving photosynthetic bacteria. The latter can be divided into three classes depending on their requirements for the assimilation of carbon dioxide. Two of these classes can use hydrogen sulfide as a substrate for growth, and produce sulfur as a by product:

$$2H_2S + CO_2 \rightarrow CH_2O + H_2O + 2S \qquad (1)$$

Third class of bacteria use organic substances like succinate as a substrate, giving fumarate as a by–product:

$$2COOHCH_2CH_2COOH + CO_2 \rightarrow CH_2O + H_2O + 2COOHCHCHCOOH \qquad (2)$$

Equations (1) and (2) with that for green plants

$$2H_2O + CO_2 \rightarrow CH_2O + H_2O + O_2$$

van Niel generalized the process of photosynthesis, and wrote a general equation:

$$2H_2A + CO_2 \rightarrow CH_2O + H_2O + 2A$$

where H_2A is an oxidizable substrate. van Niel's hypothesis has two important consequences: firstly, it gives a logical relationship between apparently diverse photosynthetic processes; secondly, it implies that water is the source of photosynthetic oxygen and that carbon dioxide is reduced to carbohydrate.

A much more direct line of evidence came from work by Hill and Scarisbrick on isolated chloroplasts, which showed that the illuminated chloroplasts would evolve oxygen only in the presence of a reducible substance such as potassium ferricyanide.

The evolution of oxygen and the reduction of the electron acceptor have identical stoichiometries, and occur in the absence of fixation of carbon dioxide. The results of this experiment indicate that the evolution of oxygen involves the oxidation of water coupled to the reduction of another substance.

In the leaf, this latter compound is carbon dioxide, but the isolation of chloroplasts frequently leads to the loss of the enzymes of the carbon dioxide fixation cycle and consequently oxygen cannot be evolved without an exogenous electron acceptor. It is only comparatively recently that techniques have been developed to isolate chloroplasts able to fix carbon dioxide at rates comparable to those found in leaves.

An obvious experiment, in the light of Hill's findings, was to use the stable, heavy isotope of oxygen (^{18}O) to discriminate between water and carbon dioxide as the source of photosynthetic oxygen. When ^{18}O became available, such an experiment was performed by Ruben, Randall, Kamen and Hyde, and the result suggested that water was the source of evolved oxygen. Such an interpretation is, however, not without problems because of the possibility of exchange reactions.

Light and dark reactions

During photosynthesis water is oxidized and carbon dioxide is reduced, but they give no indication of where in the overall process light energy intervenes to drive the reaction. However, it is—possible to show that photosynthesis consists of a combination of light–requiring reactions (the 'light reactions') and non–light –requiring reactions (the 'dark reactions'). Blackman interpreted the hyperbolic light-intensity dependence of photosynthesis as evidence for a temperature–independent photochemical reaction coupled to a 'dark' reaction which is dependent both on temperature and on the level of carbon dioxide.

Later, in the classic experiments of Emerson and Arnold here the rate of photo synthesis was measured under flashing light, it was found that the yield of photosynthesis per flash depended not only on flash intensity, but also on the time interval between flashes. The time interval between flashes which gave the maximum yield varied with temperature, but was typically 100ms, suggesting the existence of a relatively slow dark reaction which took about 100ms to be complete. The functional relationship between the 'light' and 'dark' reactions can be established by examining the requirements of the 'dark' reactions.

The 'dark' reactions comprise a complex cycle of enzyme –mediated reactions (the Calvin cycle) which catalyzes the reduction of carbon dioxide to sugar. As well as carbon dioxide the cycle requires reducing power in the form of reduced nicotinamide adenine dinucleotide phosphate (NADP) and chemical energy in the form of adenosine triphosphate (ATP). The reduced NADP (NADPH) and ATP are produced by the 'light' reactions. It is thus possible to divide a description of photosynthesis into those reactions associated with the Calvin cycle and the fixation of carbon dioxide, and those reactions which are directly driven by light.

Pigments of photosyntheis

Chloroplast is bounded by an outer double membrane or envelope, and contains a complex network of thylakoid membranes surrounded by a matrix called the stroma. The division of photosynthesis into 'light' and 'dark' reactions is reflected in this structure; the light–driven reactions all occur in or on the thylakoid membranes while the complex cycle of enzyme–mediated 'dark' reactions takes place in the stroma. The first step in the 'light' reactions of photosynthesis is the absorption of light by pigment molecules. Pigments are molecules which absorb visible, light strongly, and hence appear colored; photosynthetic pigments have extensive conjugated double bonds in the light–absorbing part of the molecule.All oxygen–evolving organisms contain chlorophyll–a (chl–a) whose structure and absorption spectrum are shown in Figure 1.

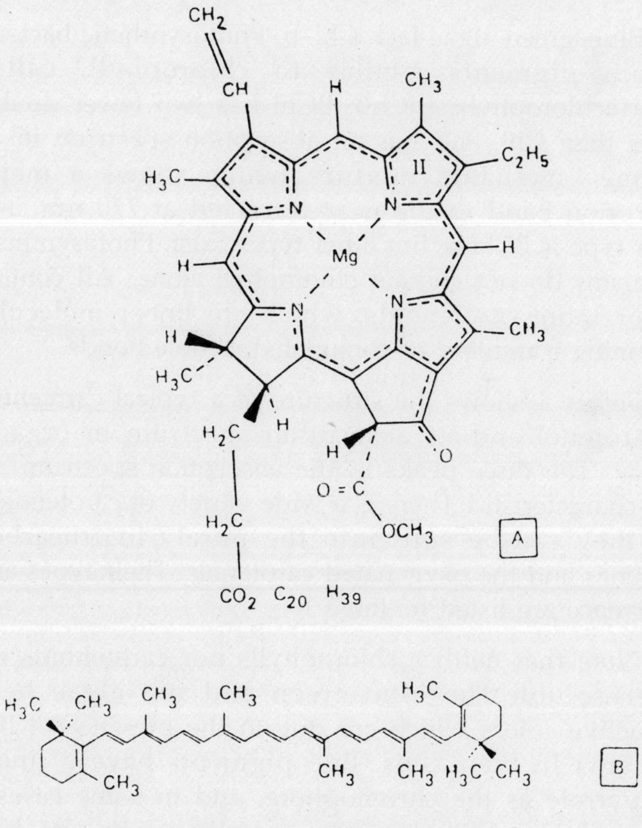

Fig. 1. Structures of chlorophyll and β-carotene.

The molecule has two distinct parts: the 'head' group consists of a cyclic tetrapyrrole with a Mg ion complexed in the center, and has extensive conjugated double bonds. The 'tail' of the molecule is a phytol side–chain which has no role in the absorption of light. Higher plants and green algae also contain chlorophyll –*b*, in which the $-CH_3$ group on ring II of Chl–*a* is replaced by a –CHO group. Two other forms of chlorophyll (Chl–c and Chl–d) are found in diatoms and some algae.

Blue–green algae lack Chl–b. Photosynthetic bacteria contain pigments similar to chlorophyll, called bacteriochlorophylls (BChl). BChl has two fewer double bonds than Chl, and has an absorption spectrum in an acetone-- methanol mixture which shows a major absorption band in the near–infra–red at 770 nm. The major type is BChl–a, but other types exist. Photosynthetic organisms do not contain chlorophyll alone. All contain one or more carotenoids, which are linear molecules containing a number of conjugated double bonds.

Figure 1 shows the structure of a typical carotenoid (β–carotene) and its absorption spectrum in organic solvent. The three peaks in the absorption spectrum are very characteristic. There is a wide variety of carotenoids, and they can be split into the purely hydrocarbon carotenes and the oxygenated carotenols. Their types and occurrence are listed in Table 1.

Note that neither chlorophylls nor carotenoids are water–soluble.The blue–green and red algae have distinctive colors which are due to the presence of bile pigments in their cells. Bile pigments have a linear tetrapyrrole as the chromophore, and in some cases a water–soluble pigment–protein complex can be extracted. The phycoerythrins absorb in the green and yellow regions of the spectrum and hence appear red; the phycocyanins appear blue. The unusual absorption characteristics of these pigments have been useful in determining the functional role of the photosynthetic pigments; the unicellular red alga *Porphyridium cruențian* has been a particularly favored organism.

Table 1
Types and occurrence of photosynthetic pigments

Pigment	*Occurrence*
Chlorophyll *a*	All oxygen-evolving organisms
Chlorophyll *b*	Higher plants and green algae
Chlorophyll *c*	Diatoms and brown algae
Chlorophyll *d*	Red algae
α–carotene	Higher plants, most algae
β–carotene	Most plants, some algae
Luteol	Higher plants, green and red algae
Violaxanthol	Higher plants
Fucoxanthol	Brown algae, diatoms
Phycoerythrin	Red algae, some blue-green algae
Phycocyanin	blue-green algae, some red algae
Allophycocyanin	Blue-green algae, red algae
Bacteriochlorophyll *a*	Purple and green bacteria
Bacteriochlorophyll *b*	Some purple bacteria

With such a variety of types of pigments it is clearly important to know how each one plays its part in the capture of light energy. They are subject to a high degree of organization, and this contributes in large measure to the efficiency of the early steps of photosynthesis.

Light harvesting

Experimental technique that led to the concept of light energy being transferred from one pigment to the next is called sensitized fluorescence. Fluorescence itself is an important property of a number of pigment molecules. If a pigment is dissolved in an organic solvent and then illuminated, the absorbed light energy has a limited number of routes via which it may be dissipated. One of these routes involves the re–emission of the light, normally at a wavelength somewhat longer than that of the exciting light, and the emitted light is called fluorescenc. The fluorescence emission spectrum is a characteristic feature of a pigment.

Experiments by Dutton, Manning and Duggar using a diatom showed that light absorbed by the carotenoid fucoxanthin excited fluorescence from Chl–a. Similarly, in the early Duysens investigated the fluorescence properties of pigments in the unicellular red alga *Porphyridium* which, contains bile pigments. He found that if the algae were illuminated with monochromatic radiation absorbed mainly by phycoerythrin then the emitted fluorescence had a fluorescence spectrum characteristic of Chl–*a*. Since Chl–*a* itself was not directly excited, this suggests that it must have been excited indirectly by excitation energy transfer from phycoerythrin. Working in this way with each pigment in turn, Duysens was able to draw up an excitation energy transfer map where light absorbed by the bile pigments is transferred to Chl–*a* through a kind of molecular bucket–brigade, and likewise Chl–*b* feeds light energy directly to Chl–*a*.

The efficiencies of energy transfer to Chl –*a* were established they are generally rather high, but sometimes the transfer from carotenoids has an efficiency of only 20–50%. Because of the time needed for excitation energy to transfer from one pigment to the next there will be a short, but finite, lag–time between excitation of B–phycoerythrin by light and the emission of fluorescence by pigments which are 'downstream' and accept the transferred excitation. Experiments designed to measure such lag–times were first performed with *Porphyridiurn* by Brody and Rabinowitch, but the most striking evidence has come with the advent of mode–locked lasers, capable of producing a train of intense pulses of light, each pulse having a duration of a few picoseconds (one picosecond = 10^{12} S; light travels approximately 0.3 mm in 1 ps). Using a mode–locked laser and a rapidly–responding photodetector, Porter and his colleagues excited one of the bile pigments in Porphyridium (B–phycoerythrin) with a pulse of wavelength 530 nm and of 6 ps duration, and monitored the fluorescence emission from B–phycoerythrin itself at 576 nm, then from R–phycocyanin (640 nm), allophycocyanin (661 nm) and Chl –*a* (685 nm).

The result is a direct confirmation of Duysens' work. Intermolecular excitation energy transfer is an important part of the function of the photosynthetic pigments, enabling wavelengths of light which Chl cannot absorb to contribute to photosynthesis.

In organic solution chlorophyll is a labile pigment. If a bright green solution of chlorophyll in acetone is left in sunlight it is bleached quickly and irreversibly to a pate straw color. This reaction requires light and oxygen, both of which are available in the leaf, and yet chlorophyll in vivo is not decolorized. Why might this be'? The precise mechanism of the bleaching of chlorophyll is not altogether clear, but it appears that the over-excitation of chlorophyll molecules by light results in the formation of so–called Chl triplet states. A triplet state is an excited state in which the spin of the excited electron has been reversed: the state is stable for several microseconds and in a reaction with molecular oxygen can form either singlet oxygen ($\Delta_g O_2$) or superoxide (O_2^-).

Both of these forms of oxygen are very reactive, and lead to irreversible damage not only to photosynthetic pigments but also to nucleic acids and other compounds. In the leaf the formation of triplet Chl appears to be followed very rapidly by the quenching of the chlorophyll triplet due to transfer of excitation energy to neighboring carotenoid molecules. This transfer leads to the formation of a harmless carotenoid triplet, whose energy is dissipated to heat.

B-phycoerythrin	
↓	70–80%
R—phycocyanin	
allophycocyanin	80-90%
↓	
Chl-*b* → Chl-*a*	

Fig. 2. Pathway of excitation transfer from the accessory pigments to Chl-a in the red alga Porphyridium cruentum.

Reactions of this type have been detected by monitoring the rapid transient absorbance changes due to carotenoid triplets in chloroplasts. The protective role of carotenoids is of vital importance: it may be the primary function of the pigments, with the light harvesting role being secondary.

In the absence of carotenoids a photosynthetic organism is at a distinct disadvantage: a mutant strain of the purple non–sulfur bacterium *Rhodopseudomonas sphaeroides* which lacks colored carotenoids must be grown strictly anaerobically since it has no mechanism to deal with bacteriochlorophyll triplet states.

In the early part of this century there was much discussion about the part played by chlorophyll in photosynthesis. A major step forward was made by Emerson and Arnold in the same series of experiments that led to the idea of 'light' and 'dark' reactions. Emerson and Arnold used neon tubes with a flash duration of about 10 μs to illuminate algal suspensions and measured manometrically the amount of oxygen evolved by a long sequence of flashes. They calculated the oxygen yield per flash and found that it was related to the energy of the flash up to a point, but that the oxygen yield saturated at relatively low flash energy.

Taking the point at which saturation has just been reached, Emerson and Arnold analyzed the algae for their chlorophyll content, and calculated the number of chlorophyll molecules associated with the evolution of one molecule of oxygen. They found the number was about 2500, and that this number was fairly constant over a range of measurements. Another ratio must be introduced here: the number of quanta that are required to evolve one molecule of oxygen.

After fierce controversy in the generally agreed that at least 8 quanta are required. Thus a fundamental association of 2500/8 $\cong$ 300 chlorophyll molecules is concerned with the processing of each quantum of light. The group of about 300 chlorophyll molecules is termed a 'photosynthetic unit'; its significance is that of the 300 chlorophyll molecules, only one 'special' Chl is concerned with the processing of light energy, while the remainder act to absorb and channel light energy to the special chlorophyll.

In this way, the effective capture cross–section of the 'special' chlorophyll is increased some 300–fold. The concept of the photosynthetic unit received further support from some calculations by Gaffron and Wohl, who estimated the time taken for oxygen evolution to begin in dark–adapted cells exposed to weak light. They calculated a long lag–time, but experimentally found a much shorter one, and suggested that chlorophyll molecules were able in some way to pool their resources by exchanging excitation energy among themselves.

The photosynthetic unit is a very important idea because it describes the high level of organization among photosynthetic pigments; however it is still unclear whether the photosynthetic unit is a statistical or a physical entity.

Nevertheless, the occurrence of excitation energy transfer between different molecules, and between like molecules, is beyond doubt. The first physical description of such excitation transfers was given by Förster who suggested that the precise mechanism would depend on the distance between donor and acceptor molecules: if R is the intermolecular distance then for large values of R (≥ 1.5 nm), the efficiency of energy transfer would depend on R^{-6}, but for small values of R (≥ 1.5 mn) the donor and acceptor molecules can 'see' each other, and the efficiency of energy transfer would depend on R^{-3}.

As well as depending on intermolecular distance, the efficiency of energy transfer is also a function (a) of the degree to which the fluorescence spectrum of the donor molecule overlaps with the absorption spectrum of the acceptor molecule and (b) of intermolecular orientation, so that energy transfer is more likely when the dipole moments of donor and acceptor molecules are aligned. This is like rotating a transistor radio to get a strong signal from the transmitter.

Förster's theories have been reviewed and extended by Knox, and some experimental studies have been reported. In physical systems of pigment mixtures, Förster's R^{-6} transfer mechanism appears to describe the intermolecular energy transfer well, but in biological systems the picture at present is not as clear.

In higher plants there appear to be only two chemically distinct forms of Chi, Chl–*a* and Chl–*b*. In the red region of the spectrum the only pigments that absorb are the chlorophylls, so that it might be expected that the absorption spectrum of the photosynthetic apparatus would be relatively simple. But in practice this is not the case: The absorption spectrum of a sample of isolated chloroplasts, and this clearly differs from the absorption spectrum of Chi in organic solvent. The difference can be understood by realizing that in the thylakoid membrane Chi does not exist in free solution, but as pigment–protein complexes. The physical environment of a pigment may affect its absorption spectrum, and the absorption spectrum of Chi–*a* is shifted to the red to a greater or lesser extent by interaction with protein, giving rise to a number of Chi – *a* species (really Chi – *a* – protein complexes) with absorption maxima in the wavelength range 650 to 720 nm.

An attempt at identification of the pigment–protein complexes can be made from the absorption spectrum by assuming that each complex has a single Gaussian absorption profile, and that the observed spectrum is the sum of a number of such profiles. Computer–assisted curve–fitting procedures can then be used to determine the number, wavelength maximum and width of each component.

The task is made easier if the sample is cooled to 77 K, since the components have reduced half–width; moreover, a fourth derivative spectrum obtained by differentiating the absorption spectrum four time with respect to wavelength can help determine the number of components. In a similar way the number of fluorescent Chi–*a*–protein components can be identified by careful study of the low–temperature fluorescence emission spectrum of chloroplasts.

Photosystems and electron transport

In the study of any photobiological process it is essential to determine the action spectrum. This spectrum shows the effectiveness or efficiency of various wavelengths of light in exciting the process, and can indicate the pigments which are involved. That the action spectrum of photosynthesis is not identical to the absorption spectrum of chlorophyll further supports the idea that other pigments absorb light which is used in driving photosynthesis. Firstly, at the short-wavelength end of the spectrum some wavelengths of light are relatively inefficient at exciting photosynthesis; this corresponds to the rather poor efficiency of excitation energy transfer between carotenoid and chlorophyll that has already been noted above. Secondly, at the long wavelength end of the spectrum, light between 680 and 720nm is also inefficient at exciting photosynthesis, although the absorption due to chlorophyll in this spectral region is still strong. This deficiency in the ability of red light to excite photosynthesis was termed the 'red drop', and might be interpreted as evidence for the existence of a long–wavelength–absorbing pool of Chl which, although it absorbs light, does not participate in photosynthesis. This interpretation is, however, inconsistent with the results of further experiments conducted by Emerson and his colleagues.

Emerson enhancement

In continuing and extending studies on the action spectra of photosynthesis, Emerson, Chalmers and Cederstrand found that the low efficiency of far–red light in exciting photosynthesis could be increased by simultaneously illuminating the sample with a second beam of light of shorter wavelength, say, at 650 nm. Not only was the low efficiency of the far–red light made up to the expected level, but the yield of photosynthesis in the combined beams was greater than the sum of the yields of the beams applied separately. This effect is known as Emerson enhancement, and can be given a numerical value, E, by the equation

$$E = \frac{\Delta O_2 \text{ (combined)} - \Delta O_2 \text{ (short - wavelength alone)}}{\Delta O_2 \text{ (long - wavelength alone)}}$$

ΔO_2 represents the rate of oxygen evolution. Emerson enhancement is as much a photobiological phenomenon as photosynthesis itself, so anEmerson enhancement is as much a photobiological phenomenon as photosynthesis itself, so an action spectrum was necessary to ascertain which pigments are involved. The results were rather complicated, but in red algae it appeared that the action spectrum for enhancement was the same as the absorption spectrum of the accessory pigments.

Thus Emerson proposed that there were two light–driven processes in photosynthesis, one sensitized by Chl–*a* and the other sensitized by the accessory pigments. It proved impossible to generalize this proposal: close examination of the action spectra of enhancement in green algae revealed that one of the species of chlorophyll–*a* itself was involved in sensitizing enhancement. This observation changed the picture: the idea of two light–driven reactions sensitized by different sets of pigments remained, but the sets of pigments were not divided as simply as originally envisioned by Emerson.

The two light–driven reactions were named by Duysens, Amesz and Kamp as Photosystem II (PSII) and Photosystem I (PSI). PSII is sensitized by several types of Chl–*a*–protein complexes and by Chl–b which is also present in a pigment–protein complex. In red algae, light absorbed by the bile pigments is preferentially transferred to PSII. In all oxygen–evolving organisms the pigments sensitizing PSII absorb only up to about 690 nm. PSI is also sensitized by Chl–a–protein complexes, but in this case the maximum wavelength of absorption is about 720nm. PSI is not strongly sensitized by Chl–*b*. If the two reactions are consecutive then the overall rate of photosynthesis will be governed by the rate–limiting step.

The explanation for the red drop now becomes clear, since at exciting wavelengths above about 690 nm PSII ceases to function, and hence the overall rate of photosynthesis decreases even though PSI still absorbs light and tries to work. The 'red drop' is eliminated when the second beam of light excites PSII and speeds up the rate-limiting step; enhancement occurs because some of this second beam excites PSI as well.

Concept of two photosystems each sensitized by a different set of pigment molecules helped to resolve a number of outstanding problems in the literature at that time. For example, several workers had studied the action spectrum of chlorophyll fluorescence in vivo, and found that in red algae light absorbed by the phycobilins was more efficient in exciting chlorophyll fluorescence than light absorbed by chlorophyll itself. The data were now interpreted as suggesting that there were two 'pools' of Chi–*a*, one which was functionally associated with the phycobilins, and which had a high fluorescence yield, and the other which was not excited by the phycobilins, and which had a low fluorescence yield.

In retrospect, it is seen that the Chi with a high fluorescence yield is associated with PSII, and that with the low fluorescence yield with PSI. Another curious observation was called 'chromatic transients'. In experiments designed to adjust the fluence rates of two wavelengths of light to produce an equal rate of photosynthesis in each, Blincks noted a transient increase in the rate of photosynthesis when one beam was switched off and the other switched on, suggesting that enhancement as described above was occurring, but that the two wavelengths of light need not be applied simultaneously to the sample.

Myers and French noted that enhancement could be elicited by flashes of different wavelengths separated by a time interval of several seconds. Taken together, these observations suggest that the photosystem excited by the one wavelength has the ability to store some intermediate which is required by the second photosystem for it to function fully.

The Z–scheme

Evidence for two light–driven photoprocesses was now clear, but the connexion between them was as yet undiscovered. In principle two methods of connexion were possible, either in series with the products of one reaction becoming the reactants of the second, or in parallel, where the two reactions proceed side by side. Cytochromes are complexes of protein and heme: the prosthetic group of the molecule is the heme, and this consists of a cyclic tetrapyrrole with an iron atom complexed in the center. Cytochromes were well known to undergo oxidation and reduction changes which were associated with changes in the cytochrome absorption spectrum. Hill had found both a *c*–type cytochrome and a b-type cytochrome and suggested that the cytochromes might act as oxidizable and reducible intermediates connecting the two photoreactions in series via electron transport.

Absorption spectrophotometry of cytochromes

The proposal of Hill and Bendall received support from the work of Duysens et al. who used a sensitive absorption – spectrophotometric technique to determine the redox state of the c–type cytochrome (cyt–f) in the red alga *Porphyridium*. The algae were illuminated with light at 680 run, which is absorbed mainly by the Chl molecules associated with PSI, and at 562 nm, which is absorbed mainly by bile pigments and transferred to PSII, while the redox state of the cytochrome was monitored by the relative absorption of a weak measuring beam at 420 nm. These observations are consistent with the series formulation or 'Z–scheme' of Hill and Bendall where the cytochrome is placed between the two photosystems. It is reduced when electrons are received from PSII and oxidized when they are passed on to PSI.

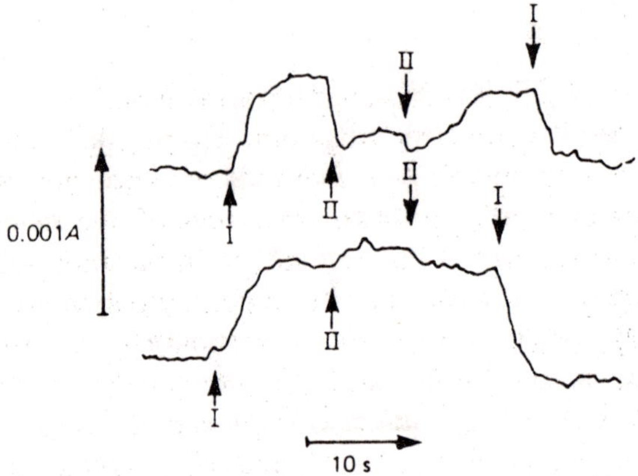

Fig. 3. Oxidation and reduction of cyt-f in prophyridium

Since the Hill and Bendall scheme was proposed, a wide variety of experimental techniques has been used further to elucidate the way in which PSII and PSI are connected. The techniques have included: (1) absorption difference spectrophotometry (as described above for cyt–f) applied to cytochromes, quinones and the copper –containing protein plastocyanin (PC); extraction and re–addition procedures, particularly for quinones electron paramagnetic resonance (epr) spectroscopy, used for those compounds which do not show marked absorbance changes when oxidized or reduced inhibitor studies and studies on mutant algae and higher plants which show impaired electron transport.

Non–cyclic electron transport

Non–cyclic electron transport involves the flow of electrons from water, through PSII and PSI in series to the terminal electron acceptor NADP. The area of the electron transport chain between water and PSII is still in doubt, but Mn is probably involved, electron transport requires Cl^- and is inhibited by a number of treatments.

On the reducing side of PSII the first stable electron acceptor is Q, probably a quinone, which can apparently be detected through its absorbance at 320 nm. Recently it has become apparent that the pool of Q is composed of two components which can be distinguished on the basis of mid–point redox potential; the low potential component (Q_{lp}) has a midpoint redox potential of around – 250 mV while the high potential component (Q_{hp}), the previously designated 0, has a mid–point potential of approximately 0 mV. Q is a single electron acceptor; the next major component in the electron transport chain is plastoquinone (PQ) which accepts two electrons and at the same time takes up two protons. The reactions of plastoquinone in vitro have been studied recently by Rich. In order for electron transport to proceed smoothly from 0 to PO there is an intermediate called R, also called B, which acts as an electron buffer, accepting two electrons from Q before handing them on to PQ. Plastoquinone also has a very important role to play in photophosphorylation.

Experiments using a mutant of the duckweed *Lemna* have indicated that electrons are passed from plastoquinone to a nonheme iron–sulfur (Fe–S) protein which is similar to that found in complex 3 of animal mitochondria, the so–called Rieske center. The Rieske center is unusual for an iron–sulfur protein, since they generally have rather negative redox potentials. In contrast to cytochromes, the iron is held to the protein via cysteine–sulfur linkages; the sulfur is called labile, since it is released by mildly acidic conditions.

The Rieske iron–sulfur center is oxidized by cytochrome f (Cyt –f) which is in its turn oxidized by plastocyanin (PC), a copper–containing protein which is the electron donor to P700. Plastocyanin appears not to be strictly necessary: in some algae non–cyclic electron transport proceeds unimpeded although PC is absent.The reducing side of PSI is rich in iron–sulfur proteins: the electron acceptors X, A and B are all thought to be bound Fe–S centers. The electrons then pass via ferredoxin–reducing substance (FRS) to a soluble Fe–S protein, ferredoxin (Fd), which is now well studied.

The terminal electron acceptor is NADP which is reduced by ferredoxin via the ferredoxin–NADP reductase. The complete non-cyclic electron transport chain has been reviewed by Malkin. At first sight it might be thought that the Z-scheme implies a fixed relationship between one PSII and one PSI that are rigidly linked by the electron transport chain. This is not so: it must be realized that the thylakoid membrane is fluid, and the PQ pool can make communication between several PSII and PSI possible. Moreover, chloroplasts may not contain equal numbers of PSII and PSI centers.

Cyclic electron transport

Electron transport does not only occur from water to NADP; there is a second 'mode' which involves neither the oxidation of water nor the reduction of NADP, and moreover only requires the excitation of photosystem 1. The pathway is called cyclic electron transport, and it was originally described by Arnon in his studies on photosynthetic phosphorylation. Cyclic electron flow is not inhibited by 3(3,4–dichlorophenyl)1, 1 – dimethyl urea (DCMU), a powerful inhibitor of non –cyclic electron flow. In addition, it may involve a cytochrome (cyt–b_6) which does not appear to be involved in non–cyclic flow.

Partial reactions

The isolation of chloroplasts normally yields a preparation that requires an exogenous electron acceptor before electron transport and the evolution of oxygen can proceed. The Hill oxidant potassium ferricyanide is one of the best known, but a variety of electron acceptors can be used which, depending on their mid–point redox potentials, will accept at different sites on the Z-scheme. Similarly, if the oxidation of water is absent or inhibited, a suitable electron donor can be provided which will enable electron transport to proceed. In this way various sections of the electron transport chain can operate. For example, PSII alone mediates a partial reaction from water to the oxidized dye dichlorophenolin-dophenol (DCIP) which is reduced to its colorless form; PSI alone mediates electron transport from reduced DCIP to a PSI acceptor such as methyl viologen. While the evolution of oxygen is normally inhibited in the latter reaction, the oxygen electrode may still be used to detect electron transport since the reduction of methyl viologen leads to the uptake of oxygen in a process known as the Mehler reaction.

Inhibitors of electron transport

One powerful technique in research into photosynthesis is the use of specific inhibitors of the electron transport chain. The best known and most widely used is DCMU, which inhibits electron transport between Q and plastoquinone.

Commercially DCMU is used as a herbicide, and a good deal of attention has been paid to the protein on the chloroplast membrane to which it binds. The binding protein can be modified by the enzyme trypsin, which can then render electron transport DCMU– insensitive. On the oxidizing side of PSII, electron transport can be inhibited by hydroxylamine, high concentrations of ammonia or methylamine, washing with high concentrations of the pH–buffer Tris, the uncoupler carbonyl cyanide 3–chlorophenyl hydrazone (CCCP), or treatment with a number of heavy metals. Between the two photosystems the plastoquinone analog 2,5–dibromo–3–methyl–6–

isopropyl–p–benzoquinone (dibromothymoquinone or DBMIB) inhibits the oxidation of plastoquinone, while plastocyanin is inhibited by poly–L–lysine, KCN or mercury. The reducing side of PSI is inhibited by disalicylidenpropane diamine (DSPD) and its water–soluble analog sulfo–DSPD.

Primary photoreactions

The reaction center, where the photoreactions take place, is clearly central to the energy transduction of photosynthesis since it mediates the first conversion of light energy to redox energy. It does this by stimulating the separation of charge, and creating a reduced electron acceptor and an oxidized electron donor. Briefly, all reaction centers contain one or two molecules of Chi that are found in a special environment which makes the Chi molecules able to use very rapidly the excitation energy for photochemical work before it can move to the next light– harvesting Chi molecule.

The reaction center chlorophyll therefore constitutes a trap for excitation energy. When the energy is captured by the reaction center chlorophyll, it is used to put the chlorophyll into a state in which giving away an electron is a favored process; the structure of the center ensures that there is a suitable electron acceptor held very close to the reaction center chlorophyll, and which can take the electron, leaving the chlorophyll oxidized. The spatial arrangement of the reaction center is also organized so that there is an electron donor in close proximity which can give an electron to the chlorophyll and return it to its former state.

Designating the electron acçeptor by A, the donor by D and the reaction center chlorcphyll by P, the sequence is then

$$DPA + \text{light energy} \rightarrow DP^{*}A \rightarrow DP^{+}A^{-} \rightarrow D^{+}PA^{-}$$

where P* represents the excited state of the chlorophyll. In purple bacteria, the ability of biochemists to isolate and purify reaction center complexes, and the ability of physicists to build light sources and detectors for spectrophotometers with picosecond time resolution has led to a good understanding of the details of charge separation in these organisms.

Photosystem I

Photosystem I is characterized by a small reversible light–induced bleaching at 700 nm, first observed by Kok. The absorption difference spectrum has the characteristics of, chlorophyll, and the pigment responsible for it was named P – 700. The bleaching was associated with an electron paramagnetic resonance (epr) signal, which indicates the appearance of an unpaired electron, and suggests that the optical bleaching was associated with photo –oxidation. This result was confirmed by use of chemical oxidants.

Careful study of the linewidth of the P700 epr signal indicates that it is consistent with P700 being a pair of Chl–a molecules, which are so closely aligned in space that functionally they act as one single molecule. However this interpretation is not unequivocal.The nature of the substance that accepts the electron when P700 is oxidized to $P700^+$ has been studied mainly by epr spectroscopy since optical spectroscopy has not proved a fruitful technique. The picture is rather complex, but it appears that the primary acceptor may also be a Chl molecule, and the first stable acceptor is a bound iron–sulfur center denoted X; the secondary acceptors appear to be two further iron–sulfur centers (A and B) close by.

Photosystem II

In some senses the reactions of photosystem II are easier to study than those of photosystem I because a wider range of experimental techniques can be applied. But in spite of this apparent advantage some of the reactions of photosystem II, particularly those associated with the oxidation of water, remain enigmatic. Like PSI, PSII exhibits a small, reversible light-induced bleaching of a chlorophyll molecule, but in this case the maximum of the bleaching is at 680 rim, and hence the reaction–center pigment of PSII is termed P680. Several epr signals are associated with PSII; the majority appear to be related to the oxidizing side of the reaction center.

The components of the reaction center, other than P680, have been difficult to identify, but it has been suggested that the primary electron acceptor is a pheophytin molecule, that is, a Chi molecule whose central Mg has been replaced by two H^+. Two phenomena associated with PSII are the emission of fluorescence from the light–harvesting chlorophylls and the oxidation of water. These will be described in turn.

Chlorophyll fluorescence

Chlorophyll fluorescence spectroscopy has considerably aided the study of PSII. It is very easy to measure, requiring only simple apparatus. Perhaps as a consequence of this simplicity the literature contains many reports on chlorophyll fluorescence: it is worth bearing in mind that even under the simplest circumstances a large number of factors come into play and influence the deceptively simple measurements.

Some of the energy absorbed by the chlorophyll molecules of PSII can be lost again as fluorescence. At room temperature in unfractionated chloroplasts, the emission is largely restricted to photosystem II; the findings of Duysens and Sweers suggested that the yield of fluorescence was under the control of the redox state of the electron acceptor Q. When Q was oxidized the fluorescence was quenched, but when Q was reduced the quenching was relieved. Indeed, Q stands for 'quencher'.

Action of Q in quenching fluorescence can be readily understood in terms of the possible fates for excitation energy in photosystem II; these are (i) re–emission as fluorescence, (ii) radiationless decay to heat, (iii) transfer to the weakly fluorescent PSI, and (iv) use in photochemistry. Giving these processes the rate constants K_f, K_h, K_t and K_p respectively, then the yield of fluorescence φ_f is given by

$$\varphi_f = \frac{K_f}{K_f + K_h + K_t + K_p A}$$

Here A is the fraction of photosystem II reaction centers that can accept an excitation, that is, centers where Q is oxidized. K_p is a relatively large rate constant when compared with the other rate constants, so φ_f is typically around 0.05. But if *A* decreases (more Q becomes reduced), φ_f will increase, because the denominator of the equation becomes smaller. Conversely, if *A* increases (more Q becomes oxidized), φ_f will decrease; hence fluorescence is an inverse measure of the yield of photochemistry. There are two extreme cases, when $A = 1$ and $A = 0$. These give

$$\varphi_o = \frac{K_f}{K_f + K_h + K_t + K_p} \quad \text{and} \quad \varphi_m = \frac{K_f}{K_f + K_h + K_t}$$

If chloroplasts are dark–adapted so that all Q becomes oxidized ($A = 1$) and then illuminated in the presence of DCMU (to prevent reduced Q becoming re–oxidized by PQ), a fluorescence induction from a low to a high level is seen. Qualitatively, the Duysens and Sweer's hypothesis accounts quite well for the low and high levels. While chlorophyll fluorescence is strictly speaking a property of PSII, the phenomenon can give information on electron transport if measurements are made in the absence of DCMU.

The chlorophyll fluorescence induction under such circumstances is called the Kautsky effect. The redox state of Q certainly plays a role in determining the fluorescence yield, but in addition a more subtle factor comes into play. The absorption spectra of PSI and PSII are not identical, and consequently it is possible to illuminate a photosynthetic system so that either PSI or PSII is over–excited, and the under–excited photosystem becomes the rate–limiting step overall.

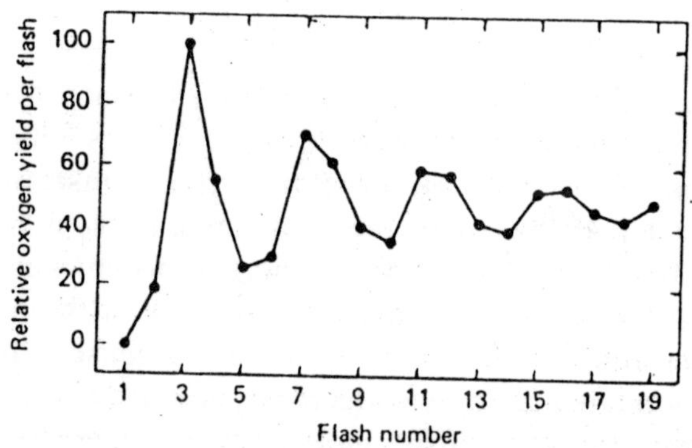

Fig. 4. The evolution of oxygen from dark-adapted Chlorella in response to a series of 5-μs flashes.

The rate constant K_t for excitation energy transfer between PSII and PSI is not fixed, but can vary in order to equalize the excitation to the two photosystems in a process known as 'spillover'. It seems likely that the complex time course of the Kautsky effect is a mixture of spillover and quenching by Q, although there is now evidence that even the carbon dioxide fixation cycle may have an influence. Chlorophyll fluorescence should be distinguished from another PSII light emission process, called delayed light emission, which is also known as luminescence or delayed fluorescence. This emission, which has an emission spectrum similar to that of 'prompt' fluorescence, is only seen after the exciting light has been extinguished. It was first observed by Strehler and Arnold, and is thought to arise as a back–reaction between metastable stored intermediates in the reaction center of PSII. The yield of emission is very low. In principle it offers a probe into the very heart of PSII, but the intensity of the emission can be modulated by a number of factors, and in practice the phenomenon is very complex.

Oxygen evolution

Evolution of oxygen from water is one of the most important reactions of photosynthesis, and yet it remains one of the least understood. A glance at the stoichiometry indicates that the oxidation of water is chemically not an easy task

$$2H_2O \rightarrow O_2 + 4H^+ + 4e^-$$

Since four electrons must be withdrawn from two molecules of water before molecular oxygen can be evolved. PSII can only withdraw one electron at a time: there are then two possibilities, either (i) four PSII reaction centers co–operate and pool their oxidizing equivalents to oxidize two molecules of water, or (ii) each PSII reaction center works independently and has a means of storing oxidizing equivalents until four are collected and can be used together. The first experimental data pertinent to distinguishing between these alternatives were obtained by Allen and Franck who found, using a sensitive phosphorescence method that in fully dark-adapted algae no oxygen was evolved by illumination with a single flash.

Similarly Joliot found that the evolution of oxygen in dark–adapted material required at least two photoacts for 'activation'. But a full answer to the question of co–operation or independence awaited a technical development, that of a rapidly– responding and sensitive oxygen electrode. The new electrode differed from the more usual Clarke–type found in many laboratories, in that the photosynthetic material was placed directly on the platinum electrode, and was not separated from it by a thin membrane.

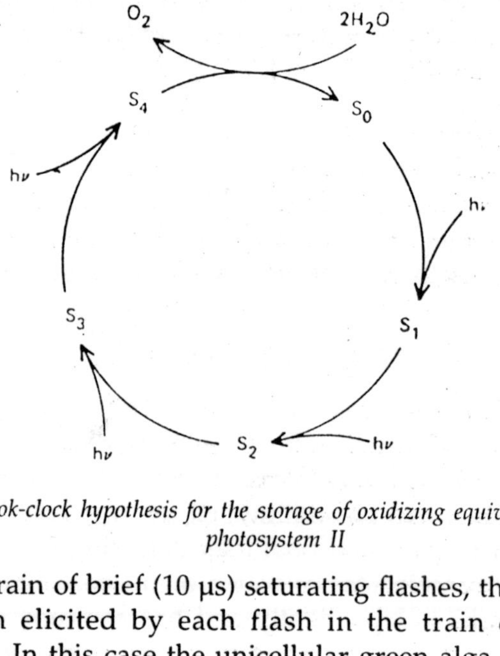

Fig. 5. Kok-clock hypothesis for the storage of oxidizing equivalents in photosystem II

Using a train of brief (10 μs) saturating flashes, the oxygen evolution elicited by each flash in the train could be observed. In this case the unicellular green alga Chlorella has been thoroughly dark–adapted before the experiment; there is no oxygen evolved on the first flash, little on the second, but a maximal amount on the third. The sequence continues with a period 4 in flash number, and gradually damps out until a steady–state level is seen on each flash. An interpretation of this phenomenon was offered by Kok, Forbush and McGloin, and incorporated the concept of charge storage by each center.

Each PSII reaction center is imagined to have a charge storage mechanism which can assume five states: S_0, S_1, S_2, S_3 and S_4 where the subscript denotes the oxidizing equivalents stored. Each absorbed photon transforms one state to the next with equal efficiency ($S_n \rightarrow S_{n+1}$); S_4 exists only transiently and reverts to S_0 after oxidizing water. The scheme needs the additional hypothesis that in the dark S_0 and S_1 are stable, and that during dark adaption the higher S–states revert mainly to S_1 and partially to S_0. There is some evidence for double hits, obtained from changing the duration of the flashes, but misses have posed a serious problem of interpretation.

The independence of centers is not quite as strict as originally conceived by Kok et al., and that the oxygen–evolving apparatus consists of a fixed photoreaction which generates an oxidizing equivalent, and a mobile protein which stores the oxidizing equivalents, Misses occur when the mobile storage protein cannot collect an oxidizing equivalent from the fixed photoreaction center. The phenomenology of flash–induced oxygen evolution says nothing about the biochemical nature of the intermediates involved. It has been known for some time that Mn is implicated in the evolution of oxygen, although not all of the pool of chloroplast Mn is essential.

Similarly, Cl^- and bicarbonate are essential co–factors for oxygen evolution. But interest has centered on Mn, since the multiplicity of oxidation states it possesses makes it plausible that it is involved in charge storage. Some use has been made of the Mn epr signal, which shows six strong peaks when it is free insolution, but these peaks disappear when it is bound. Washing chloroplasts with high concentrations of the buffer Tris is well known to inhibit the evolution of oxygen; using an epr spectrometer, Blankenship and Sauer showed that the inhibition was correlated with the release of bound Mn. Similarly treatment of chloroplasts with hydroxylamine (NH_2OH) or mild heat, both of which inhibit oxygen evolution, are associated with the release of Mn.

This circumstantial evidence was very useful, and a further step forward was made by studying the amount of Mn that could be released by mild heat treatment after each flash in a series of flashes. The amount of heat–releasable Mn shows oscillations with flash number of period 4, suggesting that it is closely associated with the oxygen–evolving mechanisms. Details of the roles Mn may play have been reviewed by Sauer. A further problem in the mechanism of oxygen evolution lies in the release of protons along with the oxidation of water. It might be thought that the protons would be released altogether at the same time as oxygen is released, but this is not the case. The sequence of proton release with respect to flash number is tricky to determine, but there is growing agreement on the sequence 1, 0, 1, 2. The experimental findings thus show a complex pattern, indicating that the oxidation of water is a multi–step process.

Biochemical investigations

The investigations described in the foregoing sections have resulted in a picture of the way that pigments and electron transport components operate together, and in particular suggest the existence in higher plants and algae of two photosystems each of which contains a reaction center and has a light–harvesting complex of molecules attached to it. It is a natural question to ask whether these functional units can be separated as physical units using biochemical techniques. There are two aims: firstly to confirm the description derived from other experiments, and secondly to make a simpler experimental material for the study of certain phenomena.

Oxygen–evolving organisms

Starting with isolated chloroplasts, a number of techniques may be used to derive preparations enriched in PSI or PSII. Boardman, Thorne and Anderson treated chloroplasts with the detergent digitonin and found that after differential centrifugation the lighter of the two fractions so obtained had enhanced PSI activity and depressed PSII activity when compared to the control, and was moreover depleted in Chl–*b*. The heavier fraction was enriched in PSII activity and Chl–*b* as compared to the control.

Subsequently it was shown that the heavy fraction was composed mainly of granal stacks and that the light fraction consisted primarily of stromal lamellae. Similar separation can be achieved using other detergents, for example Triton X– 100, or by sonication or by passage of samples through the French pressure cell. However, it can be seen that the sub–chloroplast particles obtained by these methods are not a pure preparation of one photosystem. Nevertheless, the separation provides confirmation that PSI and PSII exist. Further purification must consist in removing the pigments and electron transport components that are extraneous to the reaction centers, and it is there that the problem lies.

Many attempts have been made, but it has proved very difficult satisfactorily to fractionate the preparations further, mainly because of the instability of the extracts. Nevertheless, progress has been made towards purifying preparations of PSII. The protein composition of green plant chloroplasts has also attracted much interest thylakoid membranes are solubilized using the detergent sodium dodecyl sulfate (SDS) and the proteins separated using polyacrylamide gel electrophoresis.

The results indicate that most if not all chlorophyll exists as pigment–protein complexes and that there are at least six major pigment-proteins, although these probably represent only three distinct chlorophyll–protein complexes. These are: (i) a light–harvesting chlorophyll–protein complex which contains Chi–b, but has no light–induced oxidation – reduction activity associated with it (ii) a PSI complex, which contains P700, several forms of

Chl–a (as few as 40 Chi molecules, which include long–wavelength – absorbing forms) and a number of polypeptides; and (iii) a fraction showing some PSII activity, but not the oxidation of water. These three components which have presently been identified may well not describe the experimental data for long, since Work is progressing rapidly in this field.

Nevertheless, tentative models for the organization of two reaction center complexes and a light– harvesting complex have been proposed, including the 'tripartite' model of Butler. In addition, there have been attempts to correlate the very striking small and large particles which are apparent on freeze –fracture electron micrographs of thylakoid membranes with functional photosynthetic complexes.

Photosynthetic bacteria in marked contrast to higher plants, the isolation of functional complexes from photosynthetic bacteria was achieved several years ago. The isolation procedure is normally to break the cells, treat the exposed membranes with detergent and then use protein separation techniques to isolate a specific complex. Photochemical reaction centers devoid of light– harvesting pigments have been obtained from a number of species.

The reaction center from the purple non–sulfur bacterium Rhodopseudomonas sphaeroides contains 4 molecules of BChl and 2 of bacteriopheophytin, one or two molecules of ubiquinone and one atom of ferrous iron, together with three polypeptides of apparent molecular weights in the region of 28, 24, and 21 kD. Reaction centers of this type have been the subject of intense study, and the photochemical reactions occurring in them are now quite well understood.

In addition, the reaction center polypeptides are now the subject of sequencing and crystallographic investigations. In agreement with the concept of reaction centers and light– harvesting pigments, there are also purely light– harvesting pigment–protein complexes, the study of which is only now beginning bacterial light – harvesting pigment–protein complex has been crystallized and subjected to X–ray analysis. Seven BChl molecules are held inside a cylindrical protein, with the pigments being about 1.4 nm apart on average.

Photophosphorylation

ATP is the universal cellular energy carrier. It is formed from adenosine diphosphate (ADP) and inorganic phosphate (Pi) in oxidative and photosynthetic phosphorylation, and hydrolyzed with the consequent release of energy in a multitude of cellular reactions. The fixation of carbon dioxide within the chloroplast requires ATP; the experiments of Arnon, Allen and Whatley demonstrated that ATP could be generated by the chloroplast itself, and not, as had previously been proposed, by oxidative phosphorylation. The energy required to make the phosphate bond between ADP and Pi comes from energy lost in electron transport, whether oxidative or photosynthetic. The energy is passed via an intermediate known as the 'high–energy state' and often denoted by a so –called 'squiggle' bond, '~': it is the nature of the high–energy state that distinguishes the proposed mechanisms of phosphorylation.

Chemical hypothesis

This hypothesis, first proposed by Slater suggested that a covalent bond was formed between a component of the electron transport chain, A, and some other entity, sometimes written as X. At the start A is reduced by a reaction center, but on oxidation by the next member of the electron transport chain, B, the redox energy which is lost is captured in turning the normal bond to the '~' bond between A and X. The '~' bond is then transferred until it is the phosphate bond between ADP and Pi. In summary

$$AH_2 + X + B \rightarrow A \sim X + BH_2$$

$$A \sim X + P_i \rightarrow A + X \sim P_i$$

$$ADP + X \sim P_i \rightarrow ATP + X$$

This mechanism has been examined very closely since it does not predict transmembrane electrochemical potential gradients, nor does it give a role for closed vesicles in phosphorylation.

Chemiosmotic hypothesis

This hypothesis is proposed by Mitchell. It suggests that the high –energy state consists of a proton–concentration gradient and an electric field across the functional membrane. It is general, and applies equally well to photosynthetic and oxidative phosphorylation both in prokaryotes and eukaryotes.

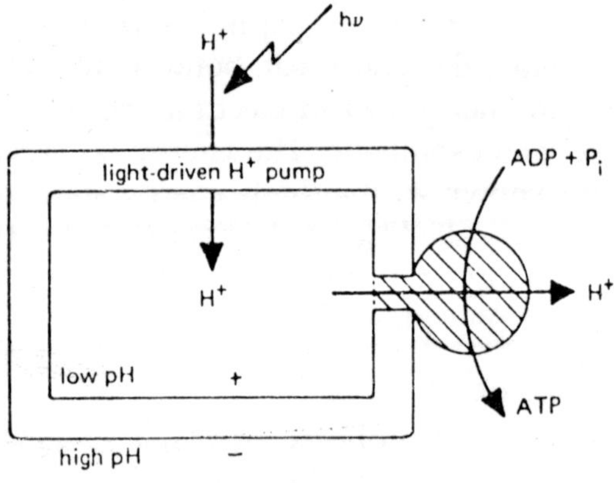

Fig. 6. Mitchell's chemiosmotic scheme for phtophosphorylation in chloroplasts.

There are five main postulates which are here applied to the chloroplast:

i) The reactions take place on the thylakoid membrane, which forms closed vesicles, and which is essentially impermeable to protons;

ii) Electron transfer from water to NADP takes place via pure electron carriers (like cytochromes) and hydrogen atom (electron plus proton) carriers; these carriers are arranged vectorially in the membrane;

iii) Electron transfer is obligatorily linked to the pumping of hydrogen ions from the stroma into the intrathylakoid space, leading to the build–up of a transmembrane proton concentration gradient (DpH) and a transmembrane field (Dy); in the chloroplast the build–up of DpH – is favoured over that of Dy by movement of counter–ions;

iv) The combination of DpH and Dy constitutes a store of energy (the high–energy state) and tends to expel protons from the intrathylakoid space;

v) There is a membrane –bound vectorial ATP–synthetase enzyme that takes the energy of the electrochemical potential gradient of protons and uses it to form ATP from ADP and P_i.

One major advantage of the chemiosmotic hypothesis was that it was open to experimental testing in a number of ways. In general, it has received good support from experiment, but there remain a number of gray areas where the precise mechanism remains in doubt. One such area is the arrangement of the proton pump; another is the functioning of the ATP–synthetase.

The proton pump

It is an essential part of the chemiosmotic hypothesis that electron transport from water to NADP and the translocation of protons from the stroma to the intrathylakoid space are obligatorily linked or coupled. That proton movements of this type take place in chloroplasts was demonstrated by Hind and Jagendorf, where the loss of protons from an essentially unbuffered medium was detected by a pH–electrode in the medium.

The increase in pH soon reached a steady state; the proton movements were reversed in the dark, and inhibited by inhibitors of electron transport, as well as uncouplers. The relationship between the electron transport chain and the proton pump clearly depends on the spatial arrangement of the electron–transport components in the thylakoid membrane. Evidence for the location of particular components can be grouped into three forms:

i) antibody agglutination or chemical labeling studies;

ii) the accessibility of components to exogenous electron donors or acceptors before and after disrupting the thylakoid membrane; and

iii) indirect studies. For example, interaction between the reaction center chlorophyll of PSI, P700, and exogenous plastocyanin occurs only after the thylakoid membrane has been disrupted, suggesting P700 is located at the interior of the thylakoid vesicle.

Similarly, antibodies raised to the protein plastocyanin cross–react only after disruption of the thylakoid, also suggesting that this component is at the interior. According to this scheme there are two modes for proton movements. In the first, plastoquinone, which when reduced is an electron–plus–proton carrier, is reduced at the stromal side of the membrane and becomes protonated there.

The reduced entity (POH_2) then diffuses to the inside of the membrane where it is oxidized, and the protons are released. This is a direct example of the obligatory coupling between electron transport and proton pumping. In the second mode, protons are taken up from the stroma as NADP is reduced, and released into the intrathylakoid space as water is oxidized, leading to an apparent movement of protons across the thylakoid membrane which is still strictly coupled to electron transport.

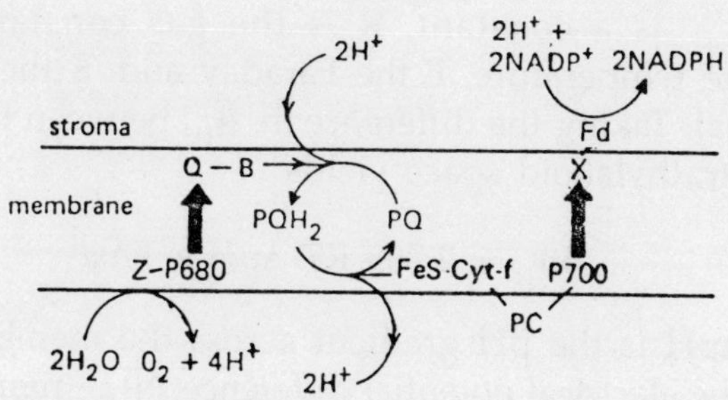

Fig. 7. Arrangement of electron-transport components in the thylakoid membrane that leads to the movement of protons from the stroma to the thylakoid space concomitant with the transport of electrons from water to NADP.

The ratio of protons translocated across the thylakoid membrane to electrons passed down the electron–transfer chain (the H^+/e ratio) has been the subject of a number of studies. In most cases the value of H+/e has been found to be 1, but there are a significant number of reports of H+/e being 2. The experiments are rather difficult to perform and interpret; more data are needed before the value of the H+/e ratio is finally settled.

High–energy state

In the chemiosmotic hypothesis the high–energy state is composed of the proton concentration gradient and the transmembrane field. Taken together, these two factors form the electrochemical potential gradient of protons, and comprise a store of energy which can be used to drive reactions in the energy–requiring direction. Thermodynamically the parameter used to describe such energy is Gibbs free energy, which is calculated from the difference in the electrochemical potentials on either side of the membrane, $\bar{\mu}_{H^+}$ and these are given by

$$\bar{\mu}_{H^+} = \bar{\mu}^*_{H^+} + RT \ln H^+ + FE$$

Here $\bar{\mu}^*_{H^+}$ is a constant, R is the gas constant, *T* the absolute temperature, F the Faraday and *E* the electrical potential. Taking the difference in $\bar{\mu}_{H^+}$ between the stroma and intrathylakoid space yields

$$\Delta\bar{\mu}_{H^+} = 2.303\ RT\ \Delta pH + F\Delta\psi$$

Here ΔpH is the pH gradient across the membrane, and Δψ is the electrical potential difference. $\Delta\bar{\mu}_{H^+}$ represents the maximum amount of free energy available to do work; the expression is usually modified by dividing through by F, and renaming $\bar{\mu}_{H^+}$ /F the proton motive force (pmf). Recognizing that 2.303 RT/F is 59 mV at 20⁰C, then

$$\text{pmf} = 59\Delta pH + Dy\ mV$$

Value of this expression is that it indicates the equal utility of ΔpH or Δψ in storing energy; thermo-dynamically the pmf can be all ΔpH and no Δψ or vice versa, or some of each. The estimation of ΔpH and Δψ can be quite tricky, and usually entails following the distribution of a radioactively labeled permeant weak acid or amine to measure ΔpH, and a labeled permeant ion in the case of Δψ. In this way, the pmf in chloroplasts was found to be mainly ΔpH, with the gradient attaining values of three pH units.

An experimental confirmation that a transmembrane pH gradient is the driving force for photophos-phorylation in chloroplasts was provided by Jagendorf and Uribe. These workers decided to eliminate the H^+-pumping part of the photophosphorylation system, and impose an artificial transmembrane pH gradient in the dark. They suspended chloroplasts in a medium of low pH, and after allowing them to equilibrate transferred them to a medium of high pH, containing ADP, Pi and Mg^{2+}.

In this way a pH gradient of the correct polarity and size was imposed across the thylakoid membrane, and ATP was formed. There is an important consequence of the buildup of a large pH gradient across the thylakoid membrane. If protons are to be pumped across the thylakoid membrane to achieve a concentration difference of around 1000–fold ($\Delta pH = 3$) then such an accumulation of positively charged ions will be accompanied by a membrane potential which is so large that it will oppose further proton influx. This will be prevented only if there is movement of counter–ions to reduce the field to a small fraction of what it otherwise would be.

Measurements of the fluxes of H^+ and other ions have been made, either with ion –sensitive electrodes, or by indirect methods and the results indicate that the movement of Mg^{2+} from the intra–thylakoid space to the stroma and Cl^- in the reverse direction may be important counter–ion fluxes. The measurement of $\Delta\psi$, at least at short times after flash excitation, can be studied very conveniently in photosynthetic organisms using spectrophotometry.

The basis of the method is that under the influence of the transmembrane field the pigments embedded in the membrane undergo a small shift in their absorption spectra, known as the electrochromic bandshift. In chloroplasts, the maximum absorption is found at 515 nm in the green region of the spectrum. On flash illumination the bandshift shows a very rapid onset (< 20 ns), concomitant with the separation of charge at the reaction centers, followed by a slow decay whose half–time (20–200ms) is related to the decay of the transmembrane field due to ion movement through the membrane. In steady–state illumination the absorption changes are much more complex because of light–induced scattering changes. Careful study of the kinetics of the onset of the electrochromic bandshift has revealed that under some circumstances the initial rapid absorption increase is followed by subsequent slower increase that may represent the electrogenic movement of charge across the membrane that occurs at sites other than the reaction centers. Observations of this type have stimulated the interest in the existence of proton motive Q–cycles in chloroplasts that has been mentioned above.

The ATP–synthetase

The vectorial enzyme which catalyzes the formation of ATP at the expense of the proton–motive force is complex: it consists of a membrane – bound proton– conducting channel termed CF_0 (coupling factor 0) and a more loosely attached catalytic section called CF_1 (coupling factor 1). If the CF_1 is removed from the membrane either by brief sonication or by incubation with ethylene diamine tetra–acetic acid (EDTA; chelates divalent cations) the phosphorylation activity of the chloroplasts is lost and the thylakoids become leaky to H^+, as judged by H^+ uptake.

In electron micrographs of suitably stained material the CF_1 protein appears as a sphere of diameter about 9 nm; it has an apparent molecular weight of 325 000 daltons. After separation from the membrane CF_1 can be reconstituted and some M^{2+}–dependent phosphorylation activity is restored. CF_1 has latent ATP hydrolysis activity which can be activated with trypsin, and which is Ca^{2+}–dependent when the enzyme is removed from the membrane. The question of how many protons efflux through the ATP–synthetase per ATP formed (the H^+/ATP ratio) can be approached in two ways. The first approach is thermodynamic, and seeks to equate the energy lost by protons as they pass through the ATP–synthetase to the energy required to phosphorylate ADP to ATP.

Suitable calculations indicate that a H^+/ATP ratio of 3 is required to account for the synthesis of ATP found under circumstances when the pmf was known. Secondly, the approach can be experimental, when H^+/ATP ratios between 2 and 4 have been found! CF_1 is composed of five subunits whose molecular weights and subunit functions are now becoming clear.

One technique to determine subunit function has been to isolate and purify each subunit and to raise antibodies to it, then to challenge the intact coupling factor with the antibodies and to see which functions are inhibited. There has been a good deal of study on the binding of nucleotides to the CF_1 and on the influence of the high–energy state on such binding, as well as the concept of conformational changes generally.

During catalysis conformational changes on the enzyme take place, detected by the energy-dependent inclusion of tritiated water into the enzyme. The soluble CF_1 contains several nucleotide–binding sites: thoroughly washed CF_1 contains 2 ATP and 1 ADP per CF_1 which exchange only slowly in the dark, but exchange in the light is more rapid and uncouplersensitive.

An elegant reconstitution study was performed by Racker and Stockenius where a light–induced proton pump and the ATPase from beef heart were inserted into artificial phospholipid vesicles. The clever part of the experiment was the choice of the proton pump; rather than try to extract and reconstitute the complex proton pumps of the chloroplast or mitochondrion, Racker and Stockenius chose the simplest possible proton pump which is the 'purple patch' from a halophylic bacterium *Halobacterium halobium*.

The purple patches and ATPase were reconstituted into the vesicles which were then illuminated, and ATP formation was found, suggesting that physical links between proton pump and ATPase were not necessary.

Uncouplers and ionophores

A brief consideration of the diagrammatic chemiosmotic scheme shows that there are three separate places where photophosphorylation may be inhibited. The first place is the proton pump which, because it is obligatorily linked to electron transport, may be inhibited by the type of electron –transport inhibitor that has already been described. Secondly, inhibition can take place at the ATP synthesizing enzyme which can either be removed from the membrane by EDTA treatment or be blocked by substances called energy–transfer inhibitors.

Table 2

The subunits of the ATP-synthetase enzyme

Designation	*Molecular weight*	*Possible stoichio-metry*	*Suggested function*
α	59 700	2	High affinity nucleotide-binding site; regulation
β	55 700	2	Low affinity nucleotide-binding site; active site
γ	36 800	1	Energy transduction from pmf to ATP
δ	19 400	1	Attachment of CF_1 to the membrane
ε	14 100	2	Inhibits ATP-hydrolase activity

The chemiosmotic hypothesis requires that the functional vesicle be impermeable to protons but a number of substances can artificially increase the proton permeability and hence wastefully dissipate the pmf. Since this breaks the link between electron transport and phosphorylation, these compounds are called 'uncouplers'. Classically, they are lipophilic weak acids which can traverse the membrane in either protonated or unprotonated form, and hence dissipate both the proton concentration gradient and the transmembrane field. Some examples are dinitrophenol

(DNP), carbonyl cyanide–*p* trifluoro-methoxy-phenyl-hydrazone (FCCP), and CCCP which has been mentioned above. There also exist ionophores which form lipophilic ionophoreion complexes and hence render membranes permeable to specific ions.

Conformational hypothesis

Events which occur within the membrane itself are important, and that the transfer of energy from electron transport to ATP formation is via forced conformational changes in the membrane or in the enzymes concerned with phosphorylation.

Fixation of carbon dioxide

Key experimental tools to solve the problem of working out the biochemical pathway for the incorporation of carbon dioxide into sugar were firstly the availability of the stable radioactive isotope of carbon, ^{14}C, and secondly the ability to separate similar substances by paper chromatography. With these tools, Calvin and his colleagues working in Berkeley in the made great strides in the understanding of the dark reactions of photosynthesis. The experiments consisted in illuminating algae in the presence of $H^{14}CO_3^-$ and after various times killing and extracting the algae in hot alcohol. The extract was then separated using two–dimensional chromatography, and radioactive compounds were identified by auto-radiography.

Calvin cycle (C–3 cycle)

Calvin and his co–workers found a general pattern in their experiments the first radioactively labeled compound was phosphoglycerate (PGA) which contains 3 carbon atoms–hence the name 'C–3 cycle'. This was produced as a result of the carboxylation of the 5–carbon compound ribulose–1.5–bisphosphate (RuBP), and the reaction was catalyzed by the most abundant protein in the world, ribulose–1,5–bisphosphate carboxylase (reaction 1).

$$3RuBP + 3CO_2 + 3H_2O \rightarrow 6PGA$$

After carboxylation, PGA is phosphorylated to diphosphoglycerate (1,3–DPGA) at the expense of a molecule of ATP.

$$6\ PGA + 6\ ATP \rightarrow 6\ (1,3 - DPGA) + 6\ ADP$$

The 1,3–DPGA is then reduced with NADPH and a proton to glyceraldehyde – 3 – phosphate (G3P), which can be regarded as the product of the cycle (reaction III)

$$6\ (1,3\text{–}DPGA) + 6\ NADPH + 6\ H^+ \rightarrow 6\ G3P + 6\ NADP+ + 6\ P_i$$

Looked at in its simplest terms, the cycle involves four reactions: carboxylation of RuBP to 2 PGA, phosphorylation of PGA to 1,3–DPGA, reduction of 1,3–DPGA to G3P and lastly the regeneration of RuBP from G3P, a reaction which also calls for chemical energy derived from ATP. The net equation of the Calvin cycle can be written

$$3CO_2 + 9ATP + 6NADPH + 6H^+ \rightarrow G3P\ (Product) + 9\ ADP + 6\ NADP^+ + 8\ P_i$$

The Calvin cycle itself is well established, attention has more recently been centered on the regulation of the cycle and on the properties of its enzymes. Regulation of the Calvin cycle has been discussed in detail by Edwards and Walker. It is important because both oxidative and reductive carbohydrate metabolism occur in the same organelle. If the Calvin cycle were not de–activated in the dark there is the possibility that ATP would be wasted in 'futile cycles', one of which might be the continuous inter–conversion of fructose–6–phosphate and fructose– 1,6–diphosphate catalyzed by the enzymes phosphofructokinase and fructose–1,6–diphosphatase. Each turn of the cycle would wastefully consume one molecule of ATP. The problem of such a 'futile cycle' could be resolved if one of the enzymes were regulated, so that both were not simultaneously catalytic.

Activation of some enzymes of the Calvin cycle by light has been known for some time. In particular, the enzymes fructose–1,6-diphosphatase, sedoheptulose–1,7-diphosphatase, NADP–glyceraldehyde–3–phosphate dehydrogenase and phosphoribulokinase are activated by light in chloroplasts in a DCMU–sensitive manner. This activation can also be achieved outside the chloroplast by reducing agents such as dithiothreitol, and has led to the suggestion that activation in vivo is due to reducing agents which are formed in the light as a result of non–cyclic electron transport.

The hypothesis has two variations; the first is proposed by Buchanan and suggests that small soluble proteins, named thioredoxin and ferredoxin – thioredoxin reductase (FTR) participate. Thioredoxin is thought to be reduced by ferredoxin via FTR, and reduced thioredoxin activates the sensitive enzyme by reducing disulfide groups on the enzyme to sulfhydryl groups.

The second variation has been proposed by Anderson and co–workers, and suggests the involvement of a membrane –bound component which is reduced directly by the electron–transport chain, and then alters the conformation of the sensitive enzyme. An alternative hypothesis for the regulation of the Calvin cycle is linked to the idea of proton movements from the stroma into the intra–thylakoid space being coupled with the movement of Mg^{2+} in the opposite direction to maintain electro-neutrality.

Since the chloroplast envelope is relatively impermeable to ions, the movements of H^+ and Mg^{2+} imply that their concentrations in the stroma will vary; both fructose– 1,6–diphosphatase and sedoheptulose–1,6–diphosphatase are sensitive to pH and Mg^{2+} concentration and it has been suggested that the movement of these ions in response to electron transport may switch on the Calvin cycle in the light. It has been argued, however, that light–induced changes in stromal Mg^{2+} levels have little significance for enzyme activation.

C–4 cycle

Labeling pattern ^{14}C found with the Calvin cycle is not found in all plants. In some, the first radioactively labeled compound contains 4 carbon atoms and plants showing this character are termed 'C–4 plants'. They have a distinguishing anatomical feature as well: the presence of a layer of bundle –sheath cells containing chloroplasts around the vascular tissue in the leaf. This feature is called Kranz anatomy, and is not seen in C–3 plants.

Moreover, there are differences in chloroplast ultrastructure between the mesophyll and bundle–sheath cells; mesophyll cell chloroplasts have normal thylakoid membranes which are stacked into grana in some regions, but lack starch grains, whereas bundle–sheath cell chloroplasts are agranal but have starch grains. There are, however, subtle differences in the ultrastructure of the bundle–sheath cell chloroplasts between the different types of C–4 plants. The distinguishing biochemical feature of C–4 plants is the original carboxylation of CO_2. The CO_2 acceptor is the 3–carbon compound phosphoenolpyruvate (PEP), and the product is the 4–carbon compound oxaloacetate (OAA).

The reaction occurs in the mesophyll cells. The fate of the OAA is of the same general pattern in all C–4 plants, but varies in the detail. Taking maize (*Zea mays*) as an example, the OAA is reduced to malate at the expense of reduced NADP; the malate so formed is then exported to the bundle–sheath cells where it is decarboxylated to pyruvate, and the CO_2 released is carboxylated by RuBP carboxylase into PGA. The pyruvate returns to the mesophyll cells where it is converted to PEP at the expense of ATP.

The C–4 cycle does not achieve any net carboxylation; it merely supplies carbon dioxide to the C – 3 cycle via a different route. It costs metabolic energy (extra ATP per CO_2), but since the K_m for CO_2 of PEP carboxylase is lower than that of RuBP carboxylase, the cycle can act as a CO_2–concentrating device. C–4 plants can be divided into three types, depending on the details of the intermediates of the cycle. The example given above, maize, is known as a malate former, since most of the radioactively-labeled carbon is found in malate, although some is found in aspartate. On the other hand there are two types of aspartate former, both of which form aspartate from OAA in mesophyll cells and then transfer it to the bundle–sheath cells, where the difference between the two types becomes apparent.

In aspartate formers of type 1, aspartate is converted back to OAA, which is then decarboxylated to PEP using a molecule of ATP as it does so. The fate of the PEP is not clear: it may be returned to the mesophyll cells, or it may participate in the return of nitrogen to the mesophyll cells. Aspartate formers of type 2 transfer aspartate to the bundle-sheath cells as described above, but the formation of OAA is followed by that of malate, which is then decarboxylated to pyruvate.

The pyruvate is transaminated to alanine, which is the route by which nitrogen returns to the mesophyll cells. The three types of C–4 plants can equally well be named according to the predominant enzyme which catalyzes the decarboxylation of the C–4 compound in the bundle–sheath cells. Malate formers have the NADP–dependent malic enzyme (NADP–ME), aspartate formers of type I have phosphoenolpyruvate carboxykinase (PEP–CK) and aspartate formers of type 2 have the NAD–dependent malic enzyme (NAD–ME). Edwards and Huber have summarized the features of each type of C–4 plant; it should be realized that the relative amount of ^{14}C found in a C–4 product is to some extent dependent on environmental and other factors.

Crassulacean acid metabolism

Crassulacean acid metabolism (CAM) is an interesting variation of carbon metabolism that occurs in succulent plants that normally inhabit dry conditions. During the night the stomata are open, and the level of malic acid in the vacuole increases, whereas during the day the stomata close and the level of malic acid falls. Carbon dioxide uptake from the atmosphere occurs predominantly during darkness.

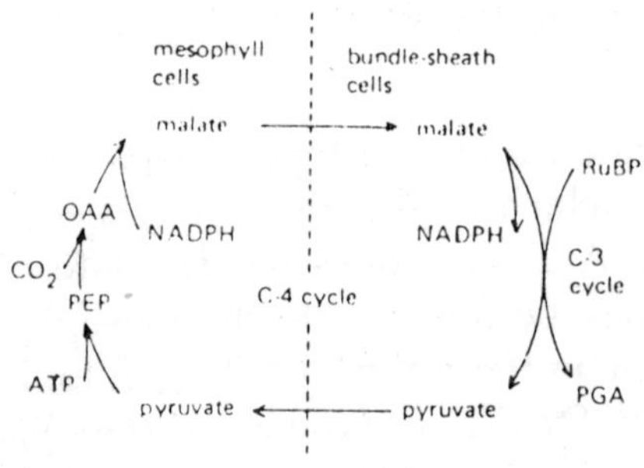

Fig. 8. C-4 cycle for fixation of carbon dioxide in a malate-forming species PEP.

CAM plants contain the enzymes of the C–4 cycle, and it appears that carboxylation of PEP to malate occurs during the night when the stomata are open, and a pool of malate is formed. During the day the stomata are closed to reduce water loss, and the malate Is decarboxylated to provide a source of CO_2 for the Calvin cycle. Hence, what the C–4 plant separates in space, the CAM plant separates in time.

Chapter 10

Basics of Photomorphogenesis

The term 'photomorphogenesis' embraces all regulatory effects of light on the development of plants, independent of photosynthesis. Correspondingly, the developmental effects of complete darkness is designated as 'skotomorphogenesis' or 'etiolation'. Both photomorphogenesis and skotomorphogenesis can be regarded as specific adaptations of the autotrophic plant which is forced to cope with greatly variable light conditions in its natural environment in order to optimize the chances of survival.

The strict separation of photomorphogenesis and photosynthesis is more than a formal logical necessity. We know that the photoreceptors used in photomorphogenesis differ from the photoreceptors of photosynthesis in almost every respect. We recall that chlorophyll, phycobilin, or rhodopsin pigments are generally present in rather large amounts in the cell (*mass pigments*) and are densely packed in membranes specialized to convert light energy into chemical energy.

In contrast, photomorphogenetic photoreceptors must be conceived as sensory pigments, the cellular level of which is generally extremely low and the function of which is not necessarily dependent on membranes. In fact, the low abundance of these photomorphogenetic pigments has prevented their physical detection until very recently and still greatly hampers their full chemical characterization, as well as the elucidation of their cellular localization and primary action mechanism. Basically, there are two different photoreceptor pigments in higher plants designed for sensing rather than converting light energy; these are known nowadays as *phytochrome* and *'cryptochrome'*. Both pigments were detected indirectly by physiological means. However, while phytochrome was spectroscopically characterized and biochemically isolated shortly after its physiological detection, *'cryptochrome'* has resisted physical and chemical identification so far, and remains a postulate derived from physiological experiments. Since the role of phytochrome in photomorphogenesis is much better established, this chapter will concentrate on the developmental functions of this pigment and refer only marginally to 'cryptochrome'. 'Cryptochrome' occurs in an even wider range of organisms, including also non-green plants such as bacteria and fungi.

Phytochrome

Biochemical and photochemical properties of phytochrome were dealt with in depth in the previous chapter and therefore only need to be reviewed very briefly. Phytochrome, a 124 000 dalton protein bearing an open-chain tetrapyrrole chromophore, represents a photochromic sensor pigment system.

Excitation of the chromophoric group by saturating amounts of light leads to the establishment of photoequilibrium between the two interconvertible forms P_r and P_{fr} which is a function of wavelength in the whole range of visible and near ultraviolet radiation.

$$\varphi_\lambda = \frac{[P_{fr}]_\lambda}{[P_{fr}]+[P_r]} = \frac{[P_{fr}]_\lambda}{[P_{tot}]}, \text{ where } P_{tot} = P_r + P_{fr}$$

where P_r (the physiologically inactive form) absorbs maximally at about 660 nm while P_{fr} (the physiologically active form) absorbs maximally at about 735 n. Thus, monochromatic light close to these two wavelengths (red and far-red light, respectively) can be conveniently used to drive the photoequilibrium of the phytochrome system either to a relative maximum of P_{fr} (φ_{660} ~ 0.8; phytochrome system maximally active) or to a state with relative low amounts of P_{fr} (e.g., φ_{730} ~ 0.01, or $\varphi_{756} < 0.001$; phytochrome system moderately active or nearly inactive). This photoreversibility of the pigment is the rationale for the well-known red/far-red reversibility of phytochrome-mediated physiological responses.

In fact, the induction of a response by a pulse of red light, the effect of which can be drastically reduced by a subsequent far-red light pulse, provides a perfect operational criterion for the identification of phytochrome as the responsible photoreceptor pigment in a particular photoresponse. However, close inspection of reveals that the phytochrome system establishes a photoequilibrium with sizable amounts of P_{fr}, and is, therefore, more or less active, in the whole range of visible and near- ultraviolet radiation. For instance, the effect of far-red light at 720 nm can be reversed by far-red light of longer wavelengths. The significance of red and far-red light in experimental work with phytochrome results mainly from the unique opportunity of reversibly converting the phytochrome system by light pulses at moderate fluence rates between extreme states of activity.

Physiological effectiveness of the phytochrome system is further complicated by several dark reactions of P_r and P_{fr}. The most important dark reactions are:

i) The formation of P_r which seems to be related to a *de novo* synthesis of the protein moiety, and

ii) The destruction of P_{fr} which seems to include the degradation of the protein moiety.

At least under certain conditions, there is evidence for several additional dark reactions, for instance a reversion of P_{fr} to P_r and a destruction of P_r. The physiological significance of these spectro-photometrically in vivo-detectable conversions is unclear; they are therefore not included in the kinetic minimal model of the phytochrome system shown in Fig. 1.

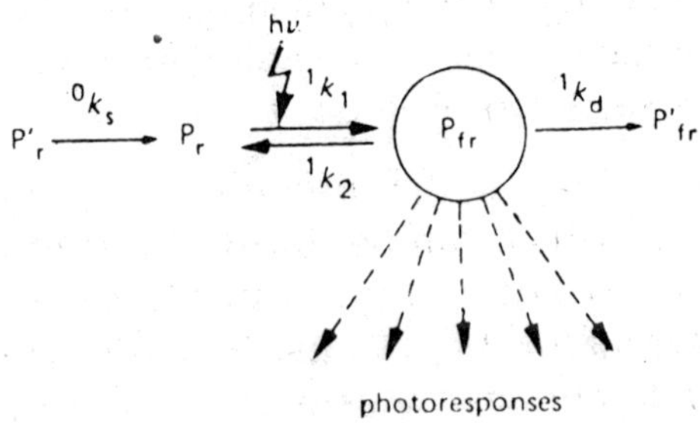

$$\frac{d[P_{tot}]}{dt} = {}^0k_s - {}^1k_d\,[P_{fr}]$$

Fig. 1. Linear kinetic model of the phytochrome system

This simple model allows the calculation of $[P_{fr}]$, (i.e., the amount of P_{fr} at time t) from $[P_{tot}]_0$ and the spectrophotometrically measurable rate constants for P_r synthesis and P_{fr} destruction. A system of light and dark reactions such as the one depicted in Fig. 1 tends to approach a stationary state with constant levels of all intermediary components whenever the rate of P_{fr} destruction ($[P_{fr}] \bullet {}^1k_d$) approaches the rate of P_r synthesis (${}^0k_s$). This takes place automatically in continuous light if the photon fluence rate is sufficient to establish the photoequilibrium $P_r \rightarrow P_{fr}$. Thus, under these conditions the amount of P_{fr} depends solely on the ratio of the two dark reactions and is, therefore, wavelength independent:

$$\frac{d[P_{tot}]}{dt} = 0,\ [P_{fr}]_{steady\ state} = \frac{{}^0k_s}{{}^1k_d}\ \text{const.}$$

Kinetic minimal model shown in Fig. 1 is sufficient to explain quantitatively, in terms of P_{fr} the induction kinetics of certain physiological responses which are controlled by a critical P_{fr} threshold. The regulation of the enzyme lipoxygenase will be used to illustrate this threshold control. In the cotyledons of the young, dark - grown mustard seedling the activity of lipoxygenase accumulates rapidly. During the period of maximal increase the enzyme activity rises by about 60%. This increase can be arrested instantaneously by the formation of a sufficient amount of P_{fr}.

Sufficient means in this case $[P_{fr}]_t/[P_{tot}]_0$ exactly ≥ 0.125, i.e., an absolute amount of P_{fr} equivalent to 1.25% or more of the amount of total phytochrome at time zero (= 36 h after sowing in these experiments). This inhibition is fully reversible. After a period of light-induced repression, the lipoxygenase activity increases again at full speed when the P_{fr} level drops below the threshold of 1.25%. Thus, the phytochrome system operates as a symmetric all-or-none switch, inhibiting enzyme formation completely at any P_{fr} level above the threshold and allowing enzyme formation with the dark-rate at any P_{fr} level below the threshold, no matter whether the threshold is surpassed in darkness (by P_{fr} destruction) or by adjustment of φ_λ with light.

Within the limits of experimental error there is no hysteresis in this regulatory system. From the extensive work on the lipoxygenase response a basic set of experiments. It is evident from the time courses of enzyme activity that the duration of enzyme repression is related to the amount of P_{fr} produced by a saturating light pulse, i.e., to the photoequilibrium adjusted by a particular wavelength up to $\varphi_{725} = 0.0125$.

Using the photometrically determined rate constant of P_{fr} destruction (${}^1k_d = 2.6 \times 10^{-4}\ s^{-1}$, $\tau_{½} = \ln 2/{}^1k_d = 45$ min) it can be shown that the P_{fr} threshold of $[P_{fr}]_\lambda / [P_{tot}]_0 = 0.0125$ is passed at each wavelength precisely when the enzyme activity abruptly starts to rise again. This type of response can obviously be used to determine physiologically active levels of phytochrome in vivo and therefore provides a critical test for the physiological significance of spectrophotometric phytochrome measurements in vivo.

A comparison of the numerical values for ${}^{1}k_d$, ${}^{0}k_s$, and φ_λ, determined either with in vivo spectrophotometry or with the lipoxygenase assay, are in amazing agreement. A clear-cut quantitative relationship between the spectrophotometrically determined dynamics of the phytochrome system and the expression of a response such as the repression of lipoxygenase increase seems to be an exception rather than a rule.

Although other threshold responses are known, the majority of phytochrome responses demonstrate a graded dependency on $[P_{fr}]$ down to very low P_{fr} levels and a complex φ_λ/effect function. Figure 2, for instance, shows the φ_λ, dependency of the anthocyanin response induced with single saturating light pulses in the mustard seedling. This response turned out to be extremely sensitive to low levels of P_{fr} but rather insensitive to changes between 5 and 80% P_{fr}.

Unfortunately this is not the only difficulty with the anthocyanin response which may be representative for many if not most phytochrome responses. Hence, the response first becomes detectable after a time which is equivalent to 3-4 half-lives of P_{fr}. There is obviously a large amount of hysteresis due to time-consuming steps within this signal response chain, which of course obliterates the $[P_{fr}]$/effect relationship on the time scale. Furthermore, phenomena such as the continuously changing responsiveness of the anthocyanin-forming cells and the stimulation of the response through a previous light treatment indicate that the signal transduction chain of the anthocyanin response-and probably of most other phytochrome responses-is considerably more sophisticated than the signal transduction chain of the lipoxygenase response.

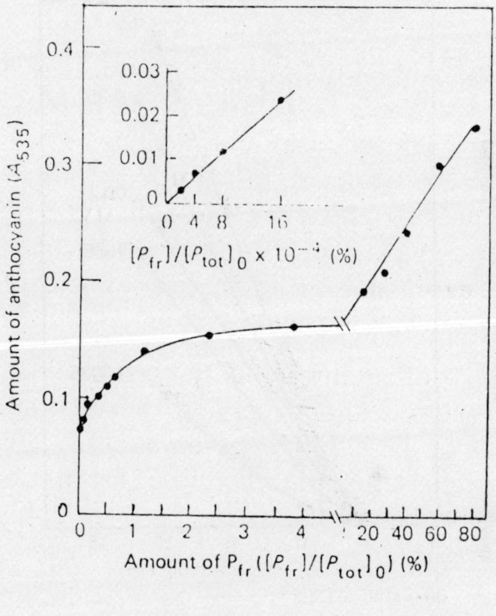

Fig. 2. Response curve of anthocyanin accumulation in mustard seedlings

While there is no reason to doubt that a clear-cut $[P_{fr}]$/ effect relationship does indeed exist in the anthocyanin and related responses, our present knowledge is far from sufficient to present a comprehensive mathematical model that would allow the calculation of time functions or $[P_{fr}]$ functions of photomorphogenetic responses from the kinetic parameters of the phytochrome system.

Theoretically, the simple phytochrome model of can also be used to calculate P_{fr} levels for continuously irradiated plants, and, in fact, these calculations yield significant results in the case of the lipoxygenase response. However, in almost every other response studied so far there are significant discrepancies between calculated P_{fr} levels and observed response levels in continuous light which are not resolved at present.

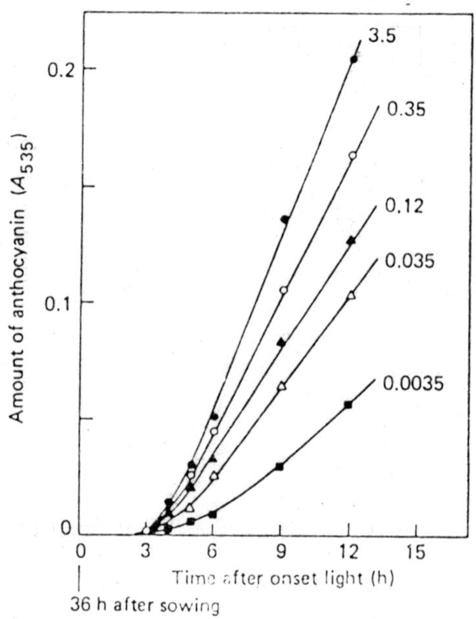

Fig. 3. High irradiance reaction in anthocyanin accumulation of mustard seelings

Briefly, these complications result from the annoying observation that in continuous light most phytochrome responses depend strongly on photon fluence rate (light intensity) although the photo-equilibrium is established after a few minutes and, therefore, the Pfr level apparently cannot increase further with increasing light absorption by P_r and P_{fr}. Since this fluence rate dependency cannot be saturated even with very bright light, the term 'high irradiance reaction has been coined for the underlying photoregulatory system. In addition to this strange fluence rate dependency, the action spectra determined under continuous light are generally characterized by a prominent peak in the far-red region and smaller peaks in the blue and near-ultraviolet region, although the linear phytochrome model predicts that $[P_{fr}]_\lambda$ = $^0k_s/^1k_d$ = constant under these conditions. Evidently this simple kinetic model of the phytochrome system fails to explain the characteristic features of the high irradiance reaction which was, therefore, attributed to an unknown blue/far-red-absorbing pigment different from phytochrome by the early investigators of these phenomena. Today this view has been replaced by the general understanding that the high irradiance reaction results from a specific mode of action of the phytochrome system, at least as far as the far-red action peak is concerned. This conclusion results from a brilliant analysis of the high irradiance response in the growth of the lettuce hypocotyl by Hartmann.

In essence, Hartmann showed that the effect of 720 nm light in the high irradiance response could be quantitatively mimicked by a mixture of 660nm and 770nm light which establishes the same φ_λ as 720 nm light, while both wavelengths are ineffective if given alone. Thus, the conclusion is unavoidable that the marked effect of continuous far-red light (around 720nm) in photomorphogenesis must somehow be related to a specific state of the phytochrome system under continuous irradiation which appears to be characterized by an optimal action at apparently low P_{fr} levels (in the range of a few per cent of P_{tot}) and by a dependency on the total number of light quanta absorbed by both P_r and P_{fr} at photoequilibrium. There have been various attempts to develop models of the phytochrome system which explain the basic features of the high irradiance reaction, a detailed discussion of which is beyond the scope of this chapter. It may suffice to mention that these models extend the minimal model by introducing at least one new element X which interacts with P_{fr} to form $P_{fr}X$.

This complex, which is thought to be the true effector molecule of the phytochrome system, is formed from P_{fr} by a rapid dark reaction and establishes a separate photoconvertible phytochrome pool $P_rX \Leftrightarrow P_{fr}X$ which can be emptied only slowly back to the free phytochrome pool $P_r \Leftrightarrow P_{fr}$ via a slow dark reaction $P_rX \rightarrow P_r + X$. Such a cyclic extension of the phytochrome model would predict, for instance, that the physiologically effective $P_{fr}X$ pool increases at the expense of the P_{fr} pool with increasing fluence rate although the overall photoequilibriumn $(P_r+P_rX) \Leftrightarrow (P_{fr}+P_{fr}X)$ remains constant. The high irradiance reaction seems to be a characteristic feature of the phytochrome system of darkgrown plants which have accumulated relatively large levels of total phytochrome.

Consequently I the high irradiance reaction is lost during prolonged light treatments which lead to a strong decrease of $[P_{tot}]$ due to P_{fr} destruction. In plants raised in continuous natural white light the characteristics of the high irradiance reaction may be largely absent. However, for the researcher interested in the molecular mechanisms of photomorphogenesis, the high irradiance reaction with continuous far-red light provides a unique experimental tool to operate the phytochrome system under steady-state conditions with respect to the effector signal. Furthermore, continuous far-red light (λ > 700nm) is inactive with respect to the photochemical protochlorophyl/chlorophyll trans-formation and, therefore, permits the study of photomorphogenesis in chlorophyll- deficient plants, avoiding interference with photosynthetic responses. These are the pragmatic reasons for the widespread use of continuous far-red light in photomorphogenetic research.

Cryptochrome

The term 'cryptochrome' describes a hypothetical photoregulatory pigment which has been characterized so far merely by its physiological action. It may well turn out in the future that 'cryptochrome' represents a variety of photomorphogenetic pigments which have in common the ability to sense light in the blue -ultraviolet spectral range and which demonstrate a superficial similarity in their absorption spectra. For the time being the operational criterion for the involvement of 'cryptochrome' in eliciting a photoresponse is an action spectrum with a peak near 370nm and three peaks or shoulders between 400 and 500 nm.

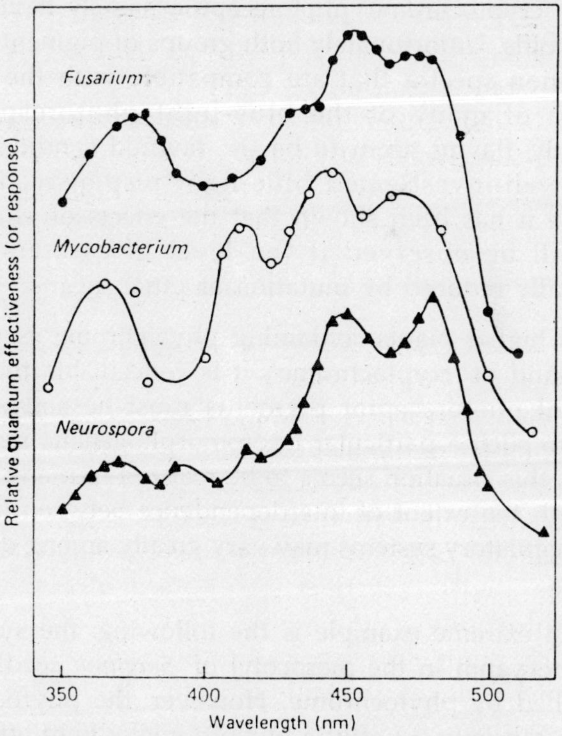

Fig. 4. Action spectra of photoinduced carotenoid biosynthesis in Fusarium equaeductuum, Mycobacterium sp., and Neurospora crassa.

In contrast to phytochrome the inductive effect of a blue -ultraviolet light treatment is not reversed by light of longer wavelengths. If the phytochrome system is also involved in a particular photoresponse, it is obviously difficult to distinguish clearly between the two photoreceptor systems since phytochrome also operates in the short wavelength region of the spectrum.

However, there are many well-established photomorphogenetic blue-light responses also in phytochrome-containing higher plants which are clearly unrelated to phytochrome, such as, for instance, the well-known phototropic response of shoots or coleoptiles. Two types of plant pigment present themselves as candidates for the 'cryptochrome' photoreceptor, namely flavins and carotenoids. Unfortunately both groups of pigments show absorption spectra that are compatible with the action spectra of many of the blue-light photoresponses. Presently, flavins seem to be the favored candidates for some well-investigated blue-light responses, mainly because it has been shown that the effects of blue light can still be observed if the level of carotenoids is drastically reduced by mutation or other means.

In higher plants containing phytochrome as well as some kind of 'cryptochrome', it is conceivable that both types of photoreceptor pigments must be activated in order to elicit a particular photomorphogenetic response. In fact, this situation seems to be realized in many plants, although the extent of interdependence between the two photoregulatory systems may vary greatly among different species.

An extreme example is the following: the synthesis of anthocyanin in the mesocotyl of *Sorghum* seedlings is controlled by phytochrome. However, the phytochrome system can only act after a blueultraviolet light effect has occurred. On the other hand, the expression of the blue -

ultraviolet effect is modified by phytochrome. It was concluded that there is an obligatory dependency between the blue-ultraviolet light and light operating through phytochrome.

It seems that somehow both pigment systems provide essential constituents of the biosynthetic chain leading to anthocyanin synthesis in this plant. Since the identification of these constituents has not been achieved so far, the neutral term coaction should be used to describe such an interdependence in photoreceptor action.

Phenomenology

Numerous morphological and physiological features that we are accustomed to regard as constitutive in a normal green plant are in fact epigenetic adaptations to the light factor produced through photomorphogenesis. A great number of developmental processes which are under photoperiodic control such as, for instance, flowering, fruit set, tuber initiation, tillering, or senescence could be included here since plants use the phytochrome system to measure the length of the daily photoperiod in order to synchronize their ontogenic cycle to the season of the year.

Many plants produce photoblastic seeds, i.e., dormant seeds which do not readily germinate unless they are treated with light. In most photoblastic seeds phytochrome can be demonstrated by red/far-red light pulse experiments to be the photoreceptor pigment responsible for induction of germination. In fact, the seeds (achehes) of Lactuca sativa have played a decisive role in the discovery of the phytochrome system. These pioneers of phytochrome research used the photoblastic Grand Rapids variety of lettuce to demonstrate for the first time the repeated red/far-red photoreversibility of phytochrome action and to determine the first true action spectra of both phytochromes. The vast amount of literature on the phenomenology of photomorphogenesis that has accumulated in the last three decades makes it quite clear that probably no higher plant can complete its life cycle without developmental control by light, mediated through the phytochrome system.

It is important to realize, however, that the controlling effects of light are generally reversible within relatively short periods of time. For instance, careful measurements of stem extension growth using electronic position transducers have shown that the growth rate of this organ can be readapted to darkness within a few minutes after inactivation of the phytochrome system.

Thus, at least during the period of rapid vegetative development, the plant can switch quickly from photomorphogenesis to skotomorphogenesis after cessation of the light stimulus. However, since the products of a previous period of development are in general rather stable, even a dramatic light-dependent modulation of the rates of morphogenetic processes will change only slowly the overall habit of a plant.

At any rate, the natural light/dark cycle in principle leads to an intermediary state of photomorphogenesis and skotomorphogenesis. Plants growing in long photoperiods (or bright light) display predominantly features of photomorphogenesis while plants growing in short photoperiods (or in the shade of a dense canopy) display features of skotomorphogenesis.

Obviously, photomorphogenesis and skotomorphogenesis are the extremes of a wide range of quantitative developmental adaptations to the light factor rather than all - or - none strategies. The quantitative interplay of light and darkness in the natural environment can be demonstrated impressively in the leaf morphogenesis of the common dandelion. The characteristic dentate leaf contour of this plant is not, as one might assume, genetically determined but is a result of photomorphogenesis. In 10 h of light per day the expression of this taxonomically decisive feature requires the continued presence of P_{fr} for some time after the end of the photoperiod. If the phytochrome system is inactivated by a far-red light pulse before the onset of the daily 14 h dark period, the dandelion leaf develops only rudimentary signs of its specific leaf morphology.

Experimental manipulation of photomorphogenesis is primarily limited to 'end-of-day' light treatments to adjust the effectiveness of the phytochrome system during the daily dark period. Because of the unavoidable interference with photosynthesis the adult green plant and its organs represent rather complex and experimentally cumbersome objects for the investigation of the mechanisms of photomorphogenesis.

Mainly for such reasons the functional aspects of photomorphogenesis have been studied almost exclusively using young, dark-grown seedlings which are capable of rapid development on the basis of their own nutritional reserves. In fact, the few days following the rupturing of the seed coat are that period of plant ontogeny in which light exerts the most dramatic effect on development. From an ecological point of view, this is quite understandable.

The ungerminated dicotyledonous seed contains an embryo, organized into embryonic organs. The embryo is provided with a limited supply of organic nutrients which can support self-sustained growth of the seedling (i.e., without access to light energy) but only for a few days. Obviously, this food store must be used most economically in order to reach the stage at which the young plant can depend on light as a source of metabolic energy. If this situation cannot be attained before exhaustion of the endogenous food reserves, the plant is irrevocably condemned to death.

It is most significant for the germinating embryo to concentrate all its developmental power on reaching the following goals: pushing the cotyledons, which are designed as prospective light- harvesting organs, through the covering layers of soil; building up the photosynthetic capacity of the cotyledons; and exposing the cotyledons maximally to the sunlight. All these developmental changes can occur in principle without further cell division. In the mustard seedling the DNA content of the cotyledons remains practically constant. In the hypocotyl the DNA content is increased by endopolyploidization; however, this takes place independently of light.

Cellular aspects

Growth, differentiation, and the integration of these processes in the multicellular plant are profoundly influenced by light acting through sensor pigments, of which phytochrome is the one most widely known. Before examining more closely the functional aspects of photomorphogenesis a few general facts need to be considered.

i) Photomorphogenesis of a plant involves numerous individual photoresponses which are highly specific for a particular organ or tissue.

ii) There is no evidence for a tissue specificity of phytochrome itself. All available data indicate that the phytochrome molecules in different parts of a plant are biochemically and spectroscopically identical.

iii) Although localization studies using immunological techniques to label phytochrome in thin sections of various plant tissues have shown that phytochrome is probably not evenly distributed within a plant, the observed quantitative differences are clearly insufficient to explain the dramatic differences in the light responsiveness of plant tissues.

Phytochrome does not carry any information with respect to the specific effects of light within a particular cell. It must be concluded, therefore, that the specificity of phytochrome responses is determined exclusively by the responding cell or cell function.

Furthermore, it must be expected that phytochrome responses depend on cell responsiveness not only qualitatively but also quantitatively. It is very unlikely that the cellular amount of P_{fr} is the decisive limiting factor determining the extent of a photoresponse under all circumstances. Rather it would be expected that the responsiveness of the cell with respect to P_{fr} can be at least as important as the P_{fr} level.

It should be emphasized, at this point, that the term 'responsiveness' does not imply any particular biochemical mechanism. It is used here with a purely descriptive meaning to summarize all relevant properties of the black box on which P_{fr} acts. These could involve the amount of P_{fr} receptor molecules as well as the biosynthetic capacity of a metabolic pathway, to mention just two possibilities. Used in the spatial meaning, cell differentiation expresses itself in the appearance of various distinctly different cell types organized inn a specific spatial pattern, notwithstanding the fact that all cells have inherited the same set of genes during cell división.

An understanding of the basic mechanisms of spatial cell differentiation is still not in sight. However, it is known that this spatial pattern includes more than just those cell features which can be detected under the microscope. For instance, this pattern also involves instructions on how to respond to external developmental factors such as light in a tissue specific manner. Two important conclusions can be drawn from an analysis of these simple drawings:

i) The individual photoresponses are strictly confined to a particular cell type that is obviously determined for expressing a particular response. We use the term 'competence' for this state of responsiveness determination.

ii) Different spatial competence patterns can be superimposed without interfering with each other.

The basic spatial program of cell differentiation provides the cells with specific sets of competences to respond to the general developmental effector phytochrome. The ultimate response, *e.g.*, synthesis of anthocyanin in the subepidermal cell layer of the mustard hypocotyl, is brought about in a two-step process.

i) establishment of the cell-specific competence toward phytochrome (= *pattern specification*);

ii) expression of this pattern through the triggering action of phytochrome (= *pattern realization*).

It is the perfectly organized interplay of these two processes that is actually causing photomorphogenesis. This situation is, of course, not unique. The theoretical concept outlined above can be extended in principle to many other developmental effectors.

Temporal pattern of competence

Theoretical framework describing photomorphogenesis in terms of space is equally applicable if we choose time as the developmental coordinate. Thus, each cell, in addition to having a specific position within the spatial pattern of competence, is further characterized by a specific temporal pattern of competence with respect to the expression of a particular photoresponse. Once again, the mustard seedling illustrates this point. As in many other plants, the embryo of mustard is essentially insensitive to photomorphogenetically active light during the periods of histodifferentiation, seed maturation and early germination although the phytochrome system operates perfectly under these conditions. It is not until some time after the young seedling has emerged from the seed coat that the responsiveness toward the phytochrome signal appears and the expression of photoresponses can be observed.

However, the decisive point is that this takes place in a sequential rather than in a synchronized manner. There is an amazingly wide range of variation for the starting points of different photoresponses. Even closely related cell functions such as the activity of neighboring enzymes of a metabolic chain can display different starting points. Using appropriate irradiation programs it can be shown that the responsiveness toward phytochrome, after surpassing the starting point, reaches an optimum and then declines to zero.

In the case of lipoxygenase repression it has been found that this enzyme expresses all-or-none kinetics with respect to the competence toward phytochrome. It enters light control suddenly at 33 h and escapes from light control suddenly at 48 h after sowing. Thus it can be concluded from the data available that, in principle, each individual photoresponse is characterized by a specific temporal competence pattern.

Photomorphogenesis of the multicellular plant therefore results from the realization by light of a precise, pre-existent time schedule of responsiveness to phytochrome. It will shortly be realized that this general statement is too simple to provide a satisfactory description of the processes involved in eliciting most phytochrome- mediated photoresponses.

There are good arguments for the notion that pattern specification as such is independent of light. In certain responses it has been found, however, that a light pretreatment, via phytochrome, can dramatically influence the quantitative effectiveness of a given amount of P_{fr} some hours later. Using the anthocyanin response of mustard, Mohr *et al.* have shown that in this case an increase as well as a reduction of phytochrome effectiveness can be produced by certain light pretreatments, even before the period of competence. This phenomenon of course greatly increases the difficulties in the interpretation of $[P_{fr}]$/response functions in continuous light.

A great deal of the surprisingly strong effect of continuous irradiation (as compared with single light pulses producing even higher P_{fr} levels) may in fact be attributed to a light-mediated potentiation of responsiveness. Mohr *et al.* have proposed that the P_{fr}-mediated change in responsiveness is caused by a change in the availability of P_{fr} receptor sites. It is, however, equally possible that the pre-irradiation mediates a change of some other metabolic factor which causes signal amplification or capacity changes at a later stage of the signal transduction chain.

Genetic aspects

Plant development can be regarded as a realization of temporal and spatial patterns which are laid down in the genetic information of DNA and finally become expressed through the catalytic action of enzymes. Factors such as light interfere with the realization of genetically determined patterns by selectively stimulating the expression of defined parts of the genetic information and inhibiting the expression of others.

From the extensive work on enzyme photoresponses there is overwhelming evidence that regulation of enzyme activities is indeed an essential element of photomorphogenetic control. Meanwhile, there are more than 50 different enzymes that have been reported to be subject to phytochrome regulation and this is probably only the tip of the iceberg.

Thus, in a general sense, everybody agrees with the statement that photomorphogenesis ultimately results from the regulation of gene activity by photosensory pigments such as phytochrome. However, there remains the decisive problem that has been a matter of dispute for many years: at which stage of gene expression, detectable on the level of enzyme activities, does phytochrome interact with the flow of genetic information?

In a first approximation this problem can be rationalized by asking the following questions: does the phytochrome signal interfere with:

i) the reading of the genetic information from the DNA (transcription), or

ii) the transformation of genetic information from the RNA code into the amino acid sequence of proteins (translation), or

iii) some *post- translational* process such as the activation or inactivation of existing enzymes?

Participation of gene activation and repression in phytochrome-mediated photomorphogenesis was suggested by Mohr and collaborators. However, mainly on the basis of experiments indicating rapid effects of phytochrome on ion transport processes and, related phenomena, there has been a strong tendency to deny the involvement of gene transcription and translation in photomorphogenetic control.

After many years of debate, the methodological progress in molecular biology has made a solution of this basic question feasible. There is now a solid body of experimental evidence that the regulation of mRNA formation and, concomitantly, the induction and repression of enzyme syntheses, are indeed essential links in the causal chain between phytochrome and developmental photoresponses. In all cases of enzyme induction thoroughly investigated so far, the rise of enzyme activity has been shown to be related to induced *de novo* synthesis of the enzyme protein. This is true for short-lived enzymes such as phenylalanine ammonia-lyase or physiologically stable enzymes such as β-amylase, both of which have been investigated intensively in the cotyledons of the mustard seedling.

The labeling of an enzyme with deuterated amino acids synthesized in the presence of deuterium oxide can be conveniently determined from the position and profile shape of isopycnically banded enzyme after density gradient centrifugation. If photomorphogenesis involves a significant increase in the synthesis and translation of mRNAs, this should also be detectable on the level of ribosomal activity. Since both the fraction of polysomes (dimers up to about 10-mers) as well as the fraction of monomers and ribosomal subunits can be quantitatively resolved, the ratio of active to inactive ribosomal material (with respect to protein synthesis) can be assessed from such gradient profiles.

It is obvious that phytochrome can induce a very rapid (lag -phase $<$ 30 min) and strong (doubling after about 1 h) increase of the relative amount of polysomes at the expense of inactive ribosomal material. The conclusion that the induced polysome formation is brought about by a rapid phytochrome-mediated increase in the supply of translatable mRNAs is supported by the fact that cordycepin, a potent inhibitor of rnRNA synthesis, completely blocks the light effect on polysome formation.

Moreover, labeling experiments with radioactive uridine have shown that the radioactivity is preferentially incorporated into the polysome fraction rather than into the ribosomal monomers and subunits during the first 30 min of the light-mediated rise of polysomes and cannot, therefore, be attributed to newly synthesized ribosomal RNA. The next step toward the gene has been pioneered by Hahlbrock and collaborators who used parsley cell cultures to demonstrate the light-induced, co-ordinated increase of the rates of PAL and chalcon synthase synthesis and the activity of the corresponding mRNAs in an in vitro translation assay.

Since then, similar results have been obtained for a variety of other phytochrome -controlled proteins, including the small subunit of ribulose bisphosphate carboxylase and the light- harvesting chlorophyll protein of the thylakoid membrane of barley and duckweed, respectively.

The synthesis of protochlorciphyllide oxidoreductase in barley leaves is repressed by phytochrome. This effect can be attributed to a repression of translatable mRNA coding for the enzyme. The demonstration of an increased level of mRNA activity, as revealed by in vitro translation, is not conclusive evidence for an increased gene transcription since there are possibly additional controlling steps in the processing and transport of mRNA which cannot be excluded by this assay.

It is therefore of significance that Link demonstrated phytochrome control of the total amount of a mRNA sequence, coding for a 35000-dalton plastid protein (of unknown function) in mustard cotyledons, using cloned DNA fragments as hybridization probes. However, it can be argued that even this experimental approach does not provide final proof for transcription control since control through mRNA degradation is still a theoretical possibility. This objection can only be ruled out by determining directly the synthesis of primary gene transcripts.

The synthesis of rRNA provides at present the only system in eukaryotic plants in which the in vivo transcription of genes of known function can be studied directly by measuring a primary transcript. The DNA sequences coding for the various RNAs of cytoplasmic ribosomes represent defined segments of a transcription unit (operon) producing a high-molecular-weight precursor rRNA which gives rise to the mature rRNAs by a sequence of specific cleavage steps. This precursor rRNA (2.5 Mdalton) can be regarded as a direct gene transcript.

The high degree of amplification of the rRNA operon (in the range of 103-104 copies per nucleus) and the specific size of the rRNA precursor molecule greatly facilitate the detection and quantification of this particular gene product by electrophoretic techniques. It has previously been shown that the accumulation of cytoplasmic rRNAs in the cotyledons of mustard seedlings can be increased through phytochrome on the basis of a constant amount of DNA.

Of course this effect, which is characterized by a relatively long lag- phase of 6 h, can hardly be related to specific photo morphogenetic responses. It may, however, be important for the adjustment of the capacity of protein synthesis during light-mediated development of the cotyledons.

The unique properties of the rRNA precursor suggested the exploration of this system for a critical test of the gene regulation hypothesis. As expected, labeling experiments with radioactive uridine in the cotyledons of mustard seedlings demonstrated a rather rapid turn-over of the rRNA precursor pool, leading to an adjustment of constant specific radioactivity about 60 min after application of the label. However, the investigation of rRNA precursor labeling before reaching the steady state provided clear evidence for a significantly enhanced rate of synthesis in far-red light-grown cotyledons as compared with the dark control.

The phytochrome effect on rRNA accumulation is in fact brought about by a change in the rate of transcription of the rRNA operon while the subsequent processing of the transcript is not influenced. This brief survey of recent achievements in the molecular analysis of photomorphogenesis strongly supports the thesis that gene transcription indeed plays a causal role in the elicitation of developmental photoresponses.

It is probably only a matter of time before the efforts presently under way in various laboratories applying cloning and hybridization techniques to genes of specific enzymes provide definite proof of the transcriptional control also in those cases where phytochrome regulation has so far been traced back only to the level of the amount or in vitro activity of the responsible mRNA. However, we should be aware that even a complete uncovering of the biochemical events elicited by the phytochrome signal at the gene level will hardly solve the mystery of how the striking specificity in time and space of phytochrome-mediated gene activation and repression can eventually be understood in molecular terms. This latter problem has not advanced beyond the level of description even during the 30 years of photomor-phogenetic research. Although the dominant role of transcriptionally mediated control in photomorphogenesis is hardly deniable, it would be premature to exclude post-transcriptional and post-translational mechanisms completely. An example warranting caution in this respect is the photoregulation of cell lengthening.

It has been found, for instance, that the response of hypocotyl growth to light can be extremely fast. The low hysteresis reasonably excludes the causal involvement of the gene regulation pathway in this response. On the contrary, there is some evidence that control of cell lengthening by light involves a control of cell wall loosening through proton excretion. Thus, the mechanism used in the photocontrol of cell growth by extension demonstrates striking similarities with the ion transport model currently discussed in auxin-mediated growth responses.

In this respect, growth by cell extension might be related more closely to those photoresponses in which the involvement of ion fluxes as the underlying cause of turgor-driven leaf movements and similar non-developmental processes is clearly established. The same may be true in seed germination which can be regarded as a special type of growth response of the young embryo at the transition from quiescence to active life.

Signal transduction

Tremendous progress in elucidation of the biochemical mechanism of single photoresponses on the enzyme level should not distract our attention from the fact that we still have no idea how the phytochrome signal is transduced from the effector molecule P_{fr} to the transciption of competent genes. Our knowledge is presently limited to indirect evidence, obtained by physiological means, which does, however, shed light on some very interesting features of this black box.

For example, it might be asked whether the switching on (or off) of the signal transduction chain of a particular photoresponse is virtually free of hysteresis, in the sense that signal transduction is rapid (and rapidly reversible) and therefore obviously involves only short-lived elements, such as Pfr itself. Alternatively, a hysteretic control system, characterized by a certain degree of back-lash, would imply at least one rather stable element which can accumulate to some extent and support the terminal response even after cessation of the phytochrome signal.

The term 'transmitter' has been used to signify such a stable intermediate of the signal transduction chain. The experimental approach for investigating the hysteresis of signal transduction chains is based on the fact that the initial phytochrome signal can be instantaneously canceled by irradiating the plant with long-wavelength far-red light, establishing a φ_λ close to zero.

Measurement of the subsequent response kinetics provides information on the lifetime of the most stable intermediate. From the after-effect of phytochrome, following the elimination of P_{fr}, we have to assume the involvement of a long-lived transmitter element in the signal chain leading to β-amylase production, but not in the signal chain leading to the production of the lyase.

Detailed time-course studies of the far-red reversibility in responses such as β-amylase induction have led to the surprising result that transmitters can accumulate under the influence of phytochrome even before the cotyledons become competent actually to produce the terminal response.

The 'coupling point', operationally defined as the start of escape from full photoreversibility, marks the time-point when the photomorphogenetic signal no longer remains restricted to the photoconvertible P_{fr} molecules but is about to pass on to a metabolic intermediate which persists even if P_{fr} is removed by the reversing light treatment. In the case of urate oxidase of mustard cotyledons this event ('coupling') can take place about 26 h before the onset of the terminal response (starting point).

In the case of peroxidase induction in mustard cotyledons the competence for transmitter formation ceases even before the starting point, leading to the seemingly paradoxical situation that the response is insensitive to P_{fr} if the plants are treated with light only from the starting point onwards. In terms of the transmitter concept, the situation for urate oxidase induction can be described as follows.

The cotyledons pass through a 40-50h period of competence toward Pfr with respect to transmitter formation. Shifted by about 26 h, a period of competence toward the transmitter with respect to urate oxidase formation follows. Except for the deviating temporal patterns, very similar conclusions can be reached from investigation of numerous other enzyme responses of the mustard cotyledon.

It is immediately evident that, depending on the extent of overlap of the two competence periods, there may be a considerable time-gap between the initial action of P_{fr} and the photomorphogenetic expression of the Pfr signal. Thus, a response may have escaped phytochrome control, and may therefore have nothing to do with the actual state of the phytochrome system, at the time when it is actually observed. The biochemical nature of transmitters is presently still unresolved. Likely candidates are, of course, mRNAs for particular proteins, at least in those responses that involve enzyme synthesis.

In the case of the mustard peroxidase response, however, it was found that the enzyme protein can be deuterium-labeled if the label is applied only during the period of transmitter formation before reaching the starting point. Moreover, inhibitors of protein synthesis are effective in suppressing the response if applied before transmitter formation, whereas the rise of enzyme activity after the starting point proves to be independent of an inhibition of protein synthesis. These results suggest that the transmitter may be identical to the inactive peroxidase protein which is made under the influence of phytochrome only before the starting point, but acquires enzymatic activity only afterwards.

In the early days of photomorphogenetic research it was generally assumed, at least implicitly, that P_{fr} participates in a single, general reaction mechanism which could in principle be discovered by analyzing any one of the many phytochrome effects. The increasing knowledge of the enormous heterogeneity of photoresponses even within the same organism, together with the apparent inability to elucidate the so-called 'primary reaction mechanism' of phytochrome, have prompted the serious questions of whether a uniform primary reaction (P_{fr} + X → $P_{fr}X$) and whether an, at least initially, uniform signal transduction mechanism of phytochrome really exist.

Indeed, an alternative concept, advocated by Mohr, proposes a multiplicity of primary reactions of P_{fr} (P_{fr} + X_1 → $P_{fr}X_1$,..., P_{fr} + X_n → $P_{fr}X_n$) to account for the conspicuous qualitative differences in the dynamics of individual photoresponses. Although a definite decision on this fundamental problem is not yet in sight, there is some relevant information from studying the hysteresis of signal transduction in the presence of full competence.

It must be recognized that the high degree of specificity of individual photoresponses even within a single cell makes it necessary to assume that the signal transduction chains, leading, e.g., to the level of enzyme synthesis, must separate somewhere in the pathway of signal transduction. This assumption is even stronger with regard to the dualism of non-developmental, rapid phytochrome responses and the developmental phytochrome responses. However, the decisive question is: at which point of the signal transduction chain does specificity come into play?

It can be concluded that probably most features of response specificity, including the direction, the timing of coupling and starting points, or the quantitative differences in [P_{fr}]/response functions, can be attributed to the later steps of the signal transduction chain, i.e., where the phytochrome signal interacts with cellular competences at or beyond the gene level.

This 'coupling reaction' is operationally defined as the first step in creating a phytochrome signal which persists for some time even after removal of P_{fr} in a reversion experiment. The time for reaching this point can be as short as 1 min in the case of the flowering response of short-day plants. In the mustard seedling, coupling times of about 2 min and more than 5 min have been reported for different responses occurring at the same time. Obviously, the phytochrome signal can be irreversibly transduced to the responding cell functions via different coupling reactions. It has been stressed repeatedly that co-operative (all-or-none) responses such as the lipoxygenase repression and non-co-operative (graded) responses such as anthocyanin accumulation are difficult to reconcile even with a common primary reaction of P_{fr}. This interpretation is true, however, only if one disregards the possibility that the initial steps of the signal transduction chain up to the coupling reaction are non-hysteretically synchronized with all changes of the active pool of P_{fr}.

There is unequivocal evidence for a multiplicity of signal transduction chains subsequent to the coupling reaction, which marks the irreversible manifestation of the phytochrome signal. This most probably takes place when the signal meets the level of cellular competence at they gene level. The steps prior to coupling, including the primary reaction of P_{fr} may be mechanistically uniform if they are precisely linked to all changes of active P_{fr}.Regardless of the ultimate answer to this question we have to accept the fact that P_{fr} represents an unspecific triggering agent of an independently running pattern of competences which decides on the specific expression and integration of photoresponses. This unavoidably leads to the disillusioning conclusion that even a full uncovering of the primary reaction(s) of P_{fr} will hardly provide significant information on the basic problems of photomorphogenesis.